Artificial Photosynthesis

RSC Foundations

For a list of titles in this series see:
rsc.li/foundations

How to obtain future titles on publication:
A standing order plan is available for this series. A standing order will bring delivery of each new volume immediately on publication.

For further information please contact:
Book Sales Department, Royal Society of Chemistry, Thomas Graham House, Science Park, Milton Road, Cambridge, CB4 0WF, UK
Telephone: +44 (0)1223 420066, Fax: +44 (0)1223 420247
Email: booksales@rsc.org
Visit our website at books.rsc.org

Artificial Photosynthesis

By

Shunichi Fukuzumi

Osaka University
Email: fukuzumi@chem.eng.osaka-u.ac.jp

RSC Foundations No. 1

Paperback ISBN: 978-1-83707-190-6
EPUB ISBN: 978-1-83767-885-3
PDF ISBN: 978-1-83707-205-7
Print ISSN: 2978-1477
Electronic ISSN: 2977-0084

A catalogue record for this book is available from the British Library

The Royal Society of Chemistry is a charity, registered in England and Wales, Number 207890, and a company incorporated in England by Royal Charter (Registered No. RC000524), registered office: Burlington House, Piccadilly, London W1J 0BA, UK, Telephone: +44 (0)20 7437 8656.

For further information see our website at www.rsc.org

For general enquiries, please contact books@rsc.org

For EU product safety enquiries, please email books@rsc.org or contact Royal Society of Chemistry Worldwide (Germany) GmbH, Römischer Hof, Unter den Linden 10, 10117 Berlin.

Preface

Artificial photosynthesis is now urgently required in order to solve global energy and environmental issues. This book focuses on molecular models of photosynthesis based on the rational design of photosynthetic systems in light of the Marcus theory of electron transfer. Natural photosynthesis is composed of two photosystems: photosystem II (PSII), where water is oxidized by plastoquinone to produce plastoquinol, and photosystem I (PA), where nicotinamide adenine dinucleotide phosphate ($NADP^+$) is reduced by plastoquinol to produce NADPH. In each photosystem, there is a photosynthetic reaction centre where photoinduced charge separation proceeds to catalyse the water oxidation in PSII and $NADP^+$ reduction in PSI. First, how the Marcus theory of electron transfer is required to understand the fundamental principles in photosynthesis is explained. Then, the rational design of the molecular models of the photosynthetic reaction centre is described to synthesize a variety of photosynthetic reaction centre models composed of organic electron donors and acceptors linked by covalent or non-covalent bonding, which undergo efficient charge separation and slow charge recombination. The efficient charge-separation step is successfully combined with the catalytic water reduction step with earth-abundant metal catalysts to develop efficient photocatalytic hydrogen evolution systems. A PSI molecular model is also developed to reduce $NAD(P)^+$ by plastoquinol analogues to produce NAD(P)H and plastoquinone analogues. A PSII molecular model is also developed to oxidize water by plastoquinone

RSC Foundations No. 1
Artificial Photosynthesis
By Shunichi Fukuzumi

Published by the Royal Society of Chemistry, www.rsc.org

analogues to evolve dioxygen. By combining PSI and PSII molecular models, the stoichiometry of photosynthesis that is $NADP^+$ reduction by water to produce NADPH is achieved as well as water splitting to hydrogen and oxygen. Hydrogen is used for catalytic hydrogenation of CO_2 to formate, CO and further reduced products such as methanol and methane. In addition, the photocatalytic oxidation of water with O_2 in the air to produce H_2O_2 is achieved together with the development of one-compartment H_2O_2 fuel cells. The photocatalytic oxidation of water with O_2 in the air was found to be enhanced significantly in seawater. Finally, the perspective of development of NADH dependent enzymatic reactions using water as an electron and proton source is discussed by combining PSI and PSII molecular models with NADH dependent enzymatic reactions to enable CO_2 reduction by water for production of solar fuels.

Shunichi Fukuzumi

Acknowledgements

The author gratefully acknowledges the contributions of their collaborators and coworkers cited in the listed references, in particular, Professor Wonwoo Nam and Professor Yong-Min Lee, and financial support by JST (CREST, SORST and ALCA) and JSPS (No. 16H02268 and 23K04686), Japan.

RSC Foundations No. 1
Artificial Photosynthesis
By Shunichi Fukuzumi

Published by the Royal Society of Chemistry, www.rsc.org

Contents

RSC Foundations No. 1
Artificial Photosynthesis
By Shunichi Fukuzumi

Published by the Royal Society of Chemistry, www.rsc.org

1 Introduction to Photosynthesis

1.1 Photosynthesis and Fossil Fuels

Photosynthesis is the process by which prokaryotic (cyanobacteria) and eukaryotic (algae and higher plants) organisms convert solar energy into chemical energy.[1] The overall chemical process of photosynthesis is the formation of carbohydrates (CH_2O) from water and CO_2 as reactants using solar energy [eqn (1.1)].[1] Photosynthesis is composed of light reactions and dark reactions (carbon dioxide fixation). The light reactions proceed *via* two photosystems, photosystem I (PSI) and photosystem II (PSII) using a set of four transmembrane protein complexes, which form the thylakoid membrane with lipids (Figure 1.1). The light reactions in PSI and PSII are started by the absorption of photons by the antenna system, in which sequential energy transfer delivers the excitation energy to PSI and PSII, both of which contain a photosynthetic reaction centre (PRC) in which a positive charge is separated from a negative charge (the so-called charge-separation process) to convert solar energy into chemical energy. Thus, the charge separation process in the PRC is the key step in photosynthesis. The principle of charge separation is discussed in relation to that in light emitting diodes in the next section.

$$CO_2 + H_2O \xrightarrow{h\nu} (CH_2O) + O_2 \tag{1.1}$$

RSC Foundations No. 1
Artificial Photosynthesis
By Shunichi Fukuzumi

Published by the Royal Society of Chemistry, www.rsc.org

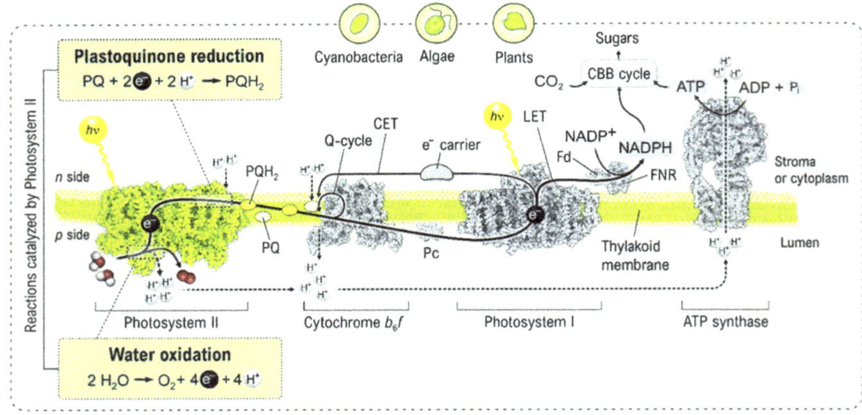

Figure 1.1 Schematic representation of photosynthesis composed of photosystem II (PSII) and photosystem I (PSI). Photoinduced electron (solid arrows) and proton (dashed arrows) transfer reactions are driven by the photosynthetic protein complexes embedded in the thylakoid membrane of oxygenic organisms (plants, algae, and cyanobacteria). The photosynthetic pigment–protein complexes, PSII, Cyt b_6f, and PSI bind the redox cofactors needed to drive the linear electron transfer (LET; long bold arrow) from H_2O $NADP^+$. The diagram also shows the cyclic electron transfer (CET) around PSI toward PQ and Cyt b_6f and then back to PSI. The reduced PQ (PQH_2) provides both the electrons and luminal protons with which NADPH and ATP are synthesized by use of ferredoxin; ferredoxin-$NADP^+$ reductase (FNR) and ATP synthase, respectively. NADPH is used for fixing CO_2 by the Calvin–Benson–Bassham (CBB) cycle. Reproduced from ref. 1, https://doi.org/10.1007/s11120-022-00991-y, under the terms of the CC BY 4.0 license, https://creativecommons.org/licenses/by/4.0/.

The combination of energy gained in PSI and PSII allows the transfer of electrons and protons from H_2O produced in PSII to the oxidized form of nicotinamide adenine dinucleotide phosphate ($NADP^+$) to produce NADPH in PSI by the so-called Z scheme [eqn (1.2) and Figure 1.1].[2] This reaction is coupled with transfer of protons from one side of the thylakoid membrane [n (negative) side; stroma in plants and algae or cytoplasm in cyanobacteria] to the other [p (positive) side or lumen].[2] By using the proton gradient, adenosine triphosphate (ATP) synthase makes ATP from adenosine diphosphate (ADP) and inorganic phosphate (Figure 1.1).[3] Thus, the solar light-driven reactions of oxygenic photosynthesis lead to the storage of solar energy in the chemical bonds of NADPH and ATP (Figure 1.1). Once, NADPH is produced from H_2O, CO_2 is reduced by NADPH to produce carbohydrates, known as the Calvin–Benson–Bassham cycle (CBB cycle), and is catalysed by the enzyme rubisco (ribulose-1,5-bisphosphate carboxylase/oxygenase) in the stroma/cytoplasm.[4]

$$2NADP^+ + 2H_2O \xrightarrow{h\nu} 2NADPH + O_2 + 2H^+ \quad (1.2)$$

The linear electron transport chain starts with PSII, where water is oxidized by PQ, leading to the production of O_2 at its electron donor side, and the reduction of plastoquinone (PQ) to plastoquinol (PQH_2) on its electron acceptor side. The overall reaction can thus be described as eqn (1.3), where the four-electron/four-proton oxidation of water (two equivalents) results in the release of four protons into the lumen (the 'p' side), whereas four protons are captured from the stromal/cytoplasmic (the 'n') side of the thylakoid membrane, accompanied by the formation of two molecules of PQH_2 (Figure 1.1). Oxidation of these two PQH_2 molecules by the cytochrome (Cyt) b_6f complex[5] leads to the release of these four protons into the lumen in addition to four protons being taken from water oxidation. Protons can be pumped by the so-called Q-cycle taking place within Cyt b_6f (Figure 1.1), and by cyclic electron transfer (CET), in which electrons are transferred back from PSI to PQ and then to Cyt b_6f. Plastocyanin (Pc), a mobile copper-containing protein, undergoes electron transfer from Cyt b_6f to PSI, where Ferredoxin-NADP$^+$ reductase (FNR) catalyses the regioselective reduction of NADP$^+$ to NADPH (1,4-reduced form) [eqn (1.3)].[6]

$$2H_2O + 2PQ + 4H^+_{stroma/cytoplasm} \xrightarrow{h\nu} O_2 + 2PQH_2 + 4H^+_{lumen} \quad (1.3)$$

$$PQH_2 + NADP^+ \xrightarrow{h\nu} PQ + NADPH + H^+ \quad (1.4)$$

Without photosynthesis, the maintenance of life on earth would be impossible. Chemical energy produced by photosynthesis performed by plants for millions of years is totally responsible for fossil fuels (coal, oil and natural gas as the solid, liquid and gas form, respectively), which were deposited in Earth's crust by sedimentation and geological processes for millions of years.[7] Fossil fuels have provided most of the energy used by humans in transportation, factories and homes. Many materials such as plastics and synthetic products used everywhere are made from fossil fuels in chemical industries. However, the year by year increasing consumption of fossil fuels by humans has caused global environmental issues such as the greenhouse effect by CO_2 emission, which may lead to serious climatic changes in the near future.[8] The consumption rate of fossil fuel may increase further at least two-fold by mid-century relative to the present because of population and economic growth. Because the consumption rate of fossil fuel resources is much faster than the production rate by natural photosynthesis, modern civilization may eventually use up fossil fuels in a few centuries. Thus, the development of artificial photosynthesis (AP) to produce solar fuels instead of fossil fuels, which should be much more efficient than natural photosynthesis, is much required to solve global energy and environmental issues.[9,10]

This book focuses on our strategy and perspective to realize *molecular* models of photosynthesis by mimicking molecular functions of PSI and PSII, mainly based on our own works for a quarter of century. Thus, heterogeneous photocatalysis in water splitting and CO_2 reduction is out of scope in this book.[11–16] The main targets are achievement of the stoichiometries of PSI [eqn (1.4)] and PSII [eqn (1.3)] using molecular models of PSI and PSII, which are then combined to achieve the overall stoichiometry [eqn (1.2)]. Once NAD(P)H is formed from H_2O and $NAD(P)^+$ with the use of solar energy, NAD(P)H can reduce CO_2 to produce a variety of value-added products. When $NAD(P)^+$ was replaced by O_2 and H^+, the combination of molecular models of PSI and PSII enabled the production of H_2O_2 and H_2, respectively. H_2O_2 can be used as a fuel in H_2O_2 fuel cells. Gaseous H_2 can be converted to liquid solar fuels by catalytic hydrogenation of CO_2. First, a variety of PRC models are developed mimicking the charge separation function of the PRC. Development of efficient photosynthetic reaction centre (PRC) models is made possible by understanding how photon energy is converted to chemical energy by controlling photoinduced electron transfer and back electron transfer. This is discussed in the next section.

1.2 Common Principles in Photosynthesis and LED

It is quite important to recognize the common principles in photosynthesis and light emitting diodes (LEDs) for the development of molecular models of photosynthesis (Scheme 1.1). Light energy is converted to chemical energy during the first stage of photosynthesis, which involves multistep photoinduced electron-transfer processes. Photoexcitation of a chromophore in the photosynthetic reaction centre results in electron transfer from the excited state of a chromophore to an adjacent electron acceptor to produce the initial charge-separated state (CS), which is rapidly converted back to the

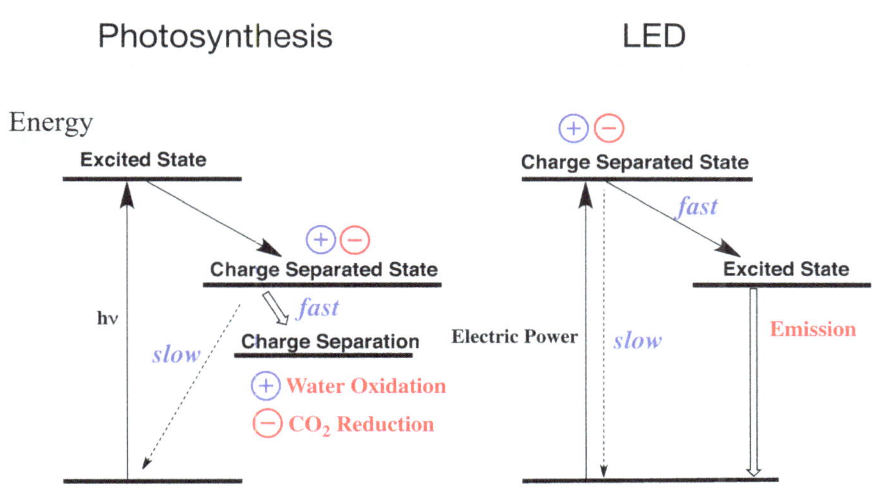

Scheme 1.1 The common principles in charge separation in photosynthesis and light emitting diodes.

ground state by back electron transfer. In such a case, light energy is converted to heat to waste energy. Thus, there should be a way to avoid rapid back electron transfer. In the photosynthetic reaction centre, there is another electron acceptor adjacent to the CS state (*vide infra*). Electron transfer from the CS state to another electron acceptor occurs to produce the next CS state in which an electron donor is more separated from the electron acceptor. The type of the adjacent electron acceptor and the distance from the initial CS state should be chosen to make the second step charge separation much faster than the back electron transfer, despite the much smaller driving force of the charge separation as compared with the large driving force of the back electron transfer to the ground state.

In contrast to the case of photosynthesis, the CS state is produced by applying the electric field and the CS energy is higher than the excited state energy of a light emitter. As in the case of photosynthesis, it is designed such that back electron transfer to the ground state with the large driving force is much slower than back electron transfer to generate the excited state with the much smaller driving force. The excited state emits the light in LED. Thus, it is extremely important to understand how the rate of electron transfer (ET) with a small driving force becomes much faster than that of ET with a much larger deriving force. This region is called the Marcus inverted region, in which the larger the ET driving force, the slower the ET rate becomes when the ET driving force is larger than that affording the fastest ET rate.[17,18] It is important to recognize that without the Marcus inverted region, neither photosynthesis nor LED would be able to proceed.

This book has focused on molecular models of photosynthesis and therefore solar fuel production with use of semiconductor catalysts is out of our scope. In order to design and construct photosynthetic model systems, the Marcus theory of electron transfer[17–19] provides a rational basis to control electron-transfer processes for solar energy conversion. Molecular model systems of photosynthesis are composed of molecular models of the photosynthetic reaction centre, photosystem I (PSI) for production of $NAD(P)^+$ and photosystem II (PSII) for water oxidation.

References

1. D. Shevela, J. F. Kern, G. Govindjee and J. Messinger, *Photosynth. Res.*, 2023, **156**, 279–307.
2. G. Govindjee, D. Shevela and L. O. Björn, *Photosynth. Res.*, 2017, **133**, 5–15.
3. W. Junge and N. Nelson, *Annu. Rev. Biochem.*, 2015, **84**, 631–657.
4. L. Gurrieri, S. Fermani, M. Zaffagnini, F. Sparla and P. Trost, *Trends Plant Sci.*, 2021, **26**, 898–912.
5. F. Tacchino, A. Succurro, O. Ebenhöh and D. Gerace, *Sci. Rep.*, 2019, **9**, 16657.
6. P. Mulo and M. Medina, *Photosynth. Res.*, 2017, **134**, 265–280.
7. B. P. Tissot and D. H. Welte, *Petroleum Formation and Occurrence*, Springer-Verlag, Berlin, 1984.
8. J. Lelieveld, K. Klingmüller, A. Pozzer, R. T. Burnett, A. Haines and V. Ramanathan, *Proc. Natl. Acad. Sci. U. S. A.*, 2019, **116**, 7192–7197.
9. T. A. Faunce, W. Lubitz, A. W. Rutherford, D. MacFarlane, G. F. Moore, P. Yang, D. G. Nocera, T. A. Moore, D. H. Gregory, S. Fukuzumi, K. B. Yoon, F. A. Armstrong, M. R. Wasielewski and S. Styring, *Energy Environ. Sci.*, 2013, **6**, 695–698.
10. S. Fukuzumi, *Joule*, 2017, **1**, 689–738.
11. Q. Wang and K. Domen, *Chem. Rev.*, 2020, **120**, 919–985.
12. S. Navalon, A. Dhakshinamoorthy, M. Alvaro, B. Ferrer and H. Garcia, *Chem. Rev.*, 2023, **123**, 445–490.
13. A. Kudo and Y. Miseki, *Chem. Soc. Rev.*, 2009, **38**, 253–278.
14. S. O. Lee, S. K. Lakheraand and K. Yong, *Adv. Energy Sustainability Res.*, 2023, **4**, 2300130.
15. X. Tao, Y. Zhao, S. Wang, C. Li and R. Li, *Chem. Soc. Rev.*, 2022, **51**, 3561–3608.
16. X.-B. Li, Z.-K. Xin, S.-G. Xia, X.-Y. Gao, C.-H. Tung and L.-Z. Wu, *Chem. Soc. Rev.*, 2020, **49**, 9028–9056.
17. R. A. Marcus, *Annu. Rev. Phys. Chem.*, 1964, **15**, 155–296.
18. R. A. Marcus, *Angew. Chem., Int. Ed. Engl.*, 1993, **32**, 1111–1121.
19. R. A. Marcus and N. Sutin, *Biochim. Biophys. Acta, Rev. Bioenerg.*, 1985, **811**, 265–322.

2 Marcus Theory of Electron Transfer

2.1 Adiabatic Electron Transfer

According to the Marcus theory of electron transfer (ET),[1–3] ET proceeds based on the Frank–Condon principle, which states that electron rearrangements occur so rapidly that nuclei can be considered as stationary until the rearrangement is complete. In such a case, the nuclear configurations of an electron donor–acceptor (D/A) pair should be the same before and after the electron transfer. As shown in Scheme 2.1, D and A form a precursor complex prior to electron transfer.[4] The nuclear configurations of the D/A pair are reorganized including solvation to reach the higher energy state, in which the nuclear configurations of the D/A pair are the same before and after ET. Then, an electron jumps from D to A to produce the $D^{\bullet+}/A^{\bullet-}$ pair, which may be reorganized to the most stable form of $D^{\bullet+}$ and $A^{\bullet-}$.

When the Gibbs energy change of ET from D to A (ΔG_{et}) is zero, the energy profiles of the D/A pair and the $D^{\bullet+}/A^{\bullet-}$ pair are shown by two parabolic curvatures (Scheme 2.2). The energy barrier of the crossing point from the reactant or product pair is one-fourth of the energy barrier for ET from D to A in the same nuclear configuration as the reactant pair, which is called reorganization energy (λ), because of the parabolic dependence of the energy on the nuclear configuration (*i.e.*, a bond distance change). The activation Gibbs energy of electron transfer form D to A is given by eqn (2.1). Typical examples of ET with

RSC Foundations No. 1
Artificial Photosynthesis
By Shunichi Fukuzumi

Published by the Royal Society of Chemistry, www.rsc.org

$\Delta G_{et} = 0$ are electron self-exchange between D and $D^{\bullet+}$ [eqn (2.2)] and that between $A^{\bullet-}$ and A [eqn (2.3)].[1]

$$\Delta G_{et}^{\neq} = \lambda/4 \tag{2.1}$$

$$D + D^{\bullet+} \rightarrow D^{\bullet+} + D \tag{2.2}$$

$$A^{\bullet-} + A \rightarrow A + A^{\bullet-} \tag{2.3}$$

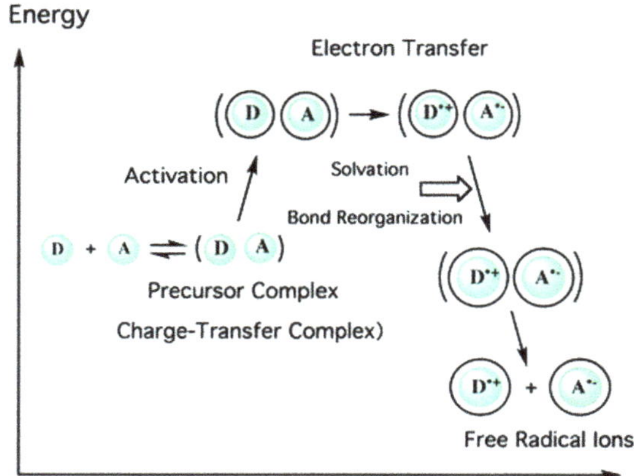

Scheme 2.1 Energy scheme of outer-sphere electron transfer based on the Frank–Condon principle.

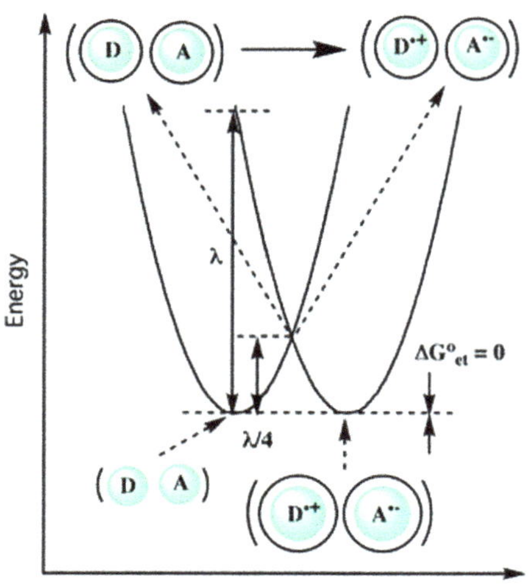

Scheme 2.2 Parabolic dependence of energy on nuclear configurations in electron transfer from D to A with $\Delta G_{et} = 0$.[4]

When ΔG_{ET} is changed, the driving force dependence of $\Delta G^{\neq}_{et}$ is given by eqn (2.4).[1]

$$\Delta G^{\neq}_{et} = (\lambda/4)[1 + (\Delta G_{et}/\lambda)]^2 \tag{2.4}$$

ΔG_{et} is given by the difference between the one-electron oxidation potential of D (E_{ox}) and the one-electron reduction potential of A (E_{red}), as shown in eqn (2.5),

$$\Delta G_{et} = e(E_{ox} - E_{red}) \tag{2.5}$$

where e is the elementary charge. When $\Delta G_{et} = -\lambda$, $\Delta G^{\neq}_{et}$ is zero and the ET rate is maximized with no need to change the nuclear configurations for ET.[1,2] When $\Delta G_{et} < -\lambda$, $\Delta G^{\neq}_{et}$ increases again with increasing ET driving force ($-\Delta G_{et}$), as shown in Scheme 2.3.[1,2] This region is called the Marcus inverted region, where ET become slower as it becomes more exergonic.[1–7] It should be emphasized again that without the Marcus inverted region, there would be no photosynthesis nor LED (*vide supra*, Scheme 1.1). The Marcus inverted region is discussed in more detail in Section 2.3.

The reaction scheme of intermolecular electron transfer from D to A is shown in Scheme 2.4.[4] The rate of disappearance of D is given by eqn (2.6), whereas the rate of formation of the precursor complex (DA) is given by eqn (2.7). By adding eqn (2.6) and eqn (2.7), eqn (2.8) is obtained. Under steady-state conditions (d[DA]/dt = 0), [DQ] is

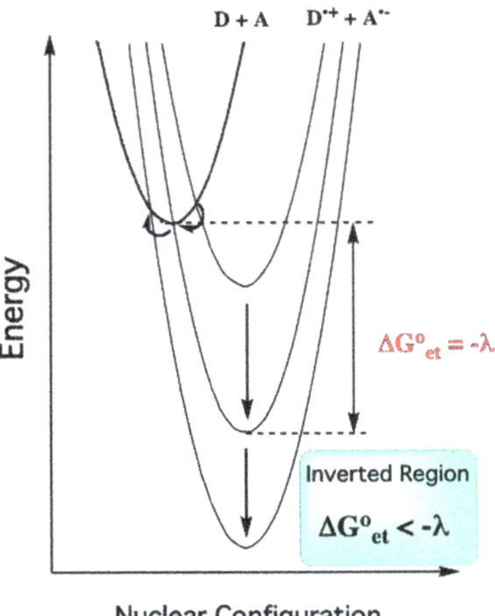

Scheme 2.3 Energy barrier dependence on the driving force of electron transfer (ΔG_{et}).

given by eqn (2.9). Then, eqn (2.10) is derived from eqn (2.8) and (2.9). Then, the relation between k_{et} and k_{ET} is given by eqn (2.11).[4]

$$d[D]/dt = -k_{12}[D][A] + k_{21}[DA] \quad (2.6)$$

$$d[DA]/dt = k_{12}[D][A] - k_{21}[DA] - k_{ET}[DA] \quad (2.7)$$

$$d([D] + [DA])/dt = -k_{ET}[DA] \quad (2.8)$$

$$[DA] = k_{12}[D][A]/(k_{ET} + k_{21}) \quad (2.9)$$

$$d[D]/dt = -k_{et}[D][A] = -\{k_{ET}k_{12}/(k_{ET} + k_{21})\}[D][A] \quad (2.10)$$

$$(k_{et})^{-1} = (k_{12})^{-1} + (k_{ET}K)^{-1} \quad (2.11)$$

where k_{12} is the diffusion rate constant to form the precursor complex (DA), k_{21} is the dissociation rate constant of DA and $K(=k_{12}/k_{21})$ is the formation constant of the precursor complex.

When $k_{ET} \gg k_{21}$, the observed rate constant (k_{et}) corresponds to the diffusion rate constant (k_{12}), eqn (2.12). When $k_{ET} \ll k_{21}$, k_{et} is given by eqn (2.12).[4] Because k_{ET} is given by eqn (2.12), k_{et} is rewritten by eqn (2.13), where $Z = k_B KT/h$ (k_B is the Boltzmann constant, T is absolute temperature and h is the Plank constant).[1] Eqn (2.16) covers the whole range of k_{ET} ($k_{ET} > k_{21}$, $k_{ET} < k_{21}$).

$$k_{ET} \gg k_{21} k_{et} = k_{12} \quad (2.12)$$

$$k_{ET} \ll k_{21} k_{et} = k_{ET}K \quad (2.13)$$

$$k_{ET} = (k_B T/h)\exp(-\lambda[1 + \Delta G_{et}/\lambda)]^2/4) \quad (2.14)$$

$$k_{et} = Z\exp(-\lambda[1 + (\Delta G_{et}/\lambda)]^2/4) \quad (2.15)$$

$$(k_{et})^{-1} = (k_{12})^{-1} + (Z\exp(-\lambda[1 + (\Delta G_{et}/\lambda)]^2/4))^{-1} \quad (2.16)$$

$$D + A \underset{k_{21}}{\overset{k_{12}}{\rightleftharpoons}} (D\,A) \xrightarrow{k_{ET}} (D^{\bullet+}\,A^{\bullet-}) \xrightarrow{fast} D^{\bullet+} + A^{\bullet-}$$

Scheme 2.4 Reaction scheme of intermolecular electron transfer from D to A.

The reorganization energy of electron transfer from D to A (λ) is given by eqn (2.17), in which λ is the average of λ_D of electron self-exchange between D and $D^{\bullet+}$ [eqn (2.18)] and λ_A between A and $A^{\bullet-}$ [eqn (2.19)].[1,2] Thus, the k_{et} value of ET from D to A can be predicted provided that E_{ox} and λ_D values of D and E_{red}, and λ_A values of A are known.

$$\lambda = (\lambda_D + \lambda_{AD})/2 \tag{2.17}$$

$$D + D^{\bullet+} \xrightarrow{\lambda_D} D^{\bullet+} + D \tag{2.18}$$

$$A + A^{\bullet-} \xrightarrow{\lambda_A} A^{\bullet-} + A \tag{2.19}$$

2.2 Nonadiabatic Electron Transfer

Electron transfer reactions in Schemes 2.2 and 2.3 are classified as either adiabatic or nonadiabatic depending on the degree of orbital interaction between D and A, as shown in Scheme 2.5, where the orbital interaction is quantified by the energy gap (ΔE) at the crossing-point nuclear configuration.[1–3] When the ΔE value is much larger than the k_BT value (*e.g.*, 0.026 eV at 298 K), two new adiabatic potential energy surfaces are produced and electron transfer along the same energy surface without jumping to the upper energy surface.[1–3] In such an adiabatic case, the transmission coefficient (κ) is unity when every incidence of the reactants reaching the crossing-point nuclear configuration leads to the formation of products with 100% probability.[4] In most intermolecular electron-transfer reactions, the ΔE value is large enough to make the electron transfer adiabatic, but small enough to be neglected in predicting the rate constant of electron transfer.[4] Such electron transfer is classified as outer-sphere electron transfer.[4] The term ‘outer-sphere’ comes from the field of coordination chemistry. Electron transfer between two coordinatively saturated metal complexes occurs without interfering with the ligand of the other metal complex.[4] The rate constant of adiabatic electron transfer (k_{ET}) is given by eqn (2.14).[4] In contrast to outer-sphere electron transfer, inner-sphere electron transfer occurs between metal complexes *via* a bridging ligand, when one of the metal complexes needs to be labile to allow the bridge to form.[4] Inner-sphere electron transfer of organic compounds may occur *via* charge-transfer complexes in which electron donor and acceptors molecules are stabilized

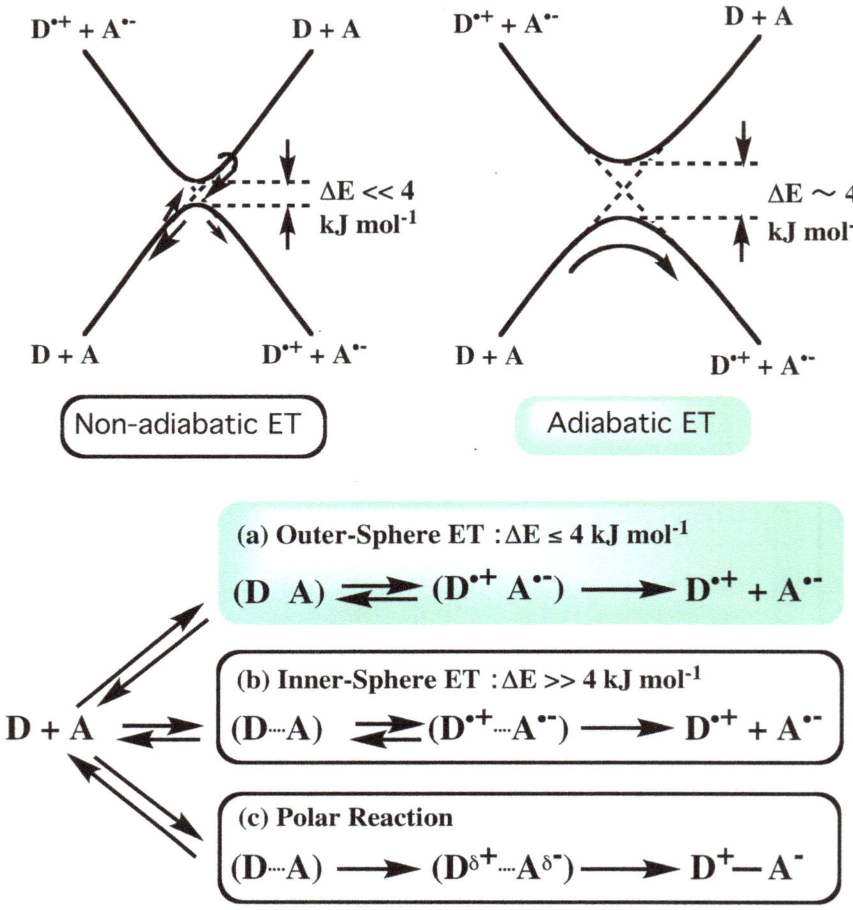

Scheme 2.5 Non-adiabatic *vs.* adiabatic ET and outer-sphere *vs.* inner-sphere ET.

by charge-transfer or hydrogen bonding interactions.[4] When interactions between donor and acceptor molecules are too strong to lead to the transition state in which only partial charge transfer occurs, such reactions are classified as polar reactions.[4]

When the electronic interaction is weak and the ΔE value is smaller than the k_BT value (*e.g.*, 0.026 eV at 298 K), the probability of electron transfer at the transition state becomes lower than unity.[4] In such a case, electron transfer does not occur along the same energy surface with certain probability jumping to the upper energy surface. Such electron transfer is classified as nonadiabatic electron transfer. Nonadiabatic electron transfer (ET) occurs in the case of intramolecular ET from an electron donor moiety (D) to an electron acceptor moiety (A) when D and A are linked with a distance (r) that is longer than the maximum distance (r_0) of adiabatic ET, which requires close contact separation between D and A. According to the Marcus theory of ET, the rate constant of nonadiabatic intramolecular ET (k_{ET}) is given by eqn (2.20),

$$k_{ET} = \left(\frac{4\pi^3}{h^2\lambda k_R T}\right)^{1/2} V^2 \exp\left[-\frac{(\Delta G_{ET} + \lambda)^2}{4\lambda k_B T}\right] \tag{2.20}$$

where V is the electronic coupling matrix element. ΔG_{ET} is the free energy change of intramolecular ET, which is virtually the same as that of eqn (2.5), and λ is the reorganization energy of electron transfer.[3] In eqn (2.20), V^2 is given as a function of distance of D–A ($r - r_0$) under the conditions of $r > r_0$, as shown in eqn (2.21).[3] When $r < r_0$, k_{ET} is given by eqn (2.14) for adiabatic ET.[3]

$$V^2 = V_0{}^2 \exp[-\beta(r - r_0)] \tag{2.21}$$

2.3 Marcus Inverted Region

According to eqn (2.20) and (2.21), the k_{ET} value decreases exponentially with increasing r of D–A. Such long-range slow thermal ET would be achieved if D or A molecules are encapsulated in a large inert environment which prohibits close access of the other molecule. Such long-range slow ET was reported by using a Y-type zeolite supercage in which electron acceptor molecules are encapsulated by cation exchange reactions.[8] A metal cation complex, $[Fe(bpy)_3]^{3+}$ (bpy = 2,2′-bipyridine) was incorporated into the NaY zeolite supercage by ship-in-bottle synthesis (Figure 2.1).[8] The size of $[Fe(bpy)_3]^{3+}$ (12 Å diameter) makes for a secure fit inside the zeolite supercage (13 Å). ET from ferrocene $[Fe(C_5H_5)_2]$ to $[Fe(bpy)_3]^{3+}$–zeolite Y suspended in acetonitrile (MeCN) resulted in a gradual increase in absorption at 520 nm due to $[Fe(bpy)_3]^{2+}$ produced by ET from $Fe(C_5H_5)_2$ to $[Fe(bpy)_3]^{3+}$ inside the zeolite.[8] The Gibbs energy change of ET from $Fe(C_5H_5)_2$ (E_{ox} *vs.* SCE = 0.37 V) to $[Fe(bpy)_3]^{3+}$ (E_{red} *vs.* SCE = 1.06 V) in MeCN is highly exergonic ($\Delta G_{et} = -0.69$ eV).[8] Thus, intermolecular ET from $Fe(C_5H_5)_2$ to $[Fe(bpy)_3]^{3+}$ occurred very rapidly within 100 ms upon mixing $Fe(C_5H_5)_2$ and $[Fe(bpy)_3]^{3+}$.[8] In sharp contrast, the observed rate of ET from $Fe(C_5H_5)_2$ to $[Fe(bpy)_3]^{3+}$ inside

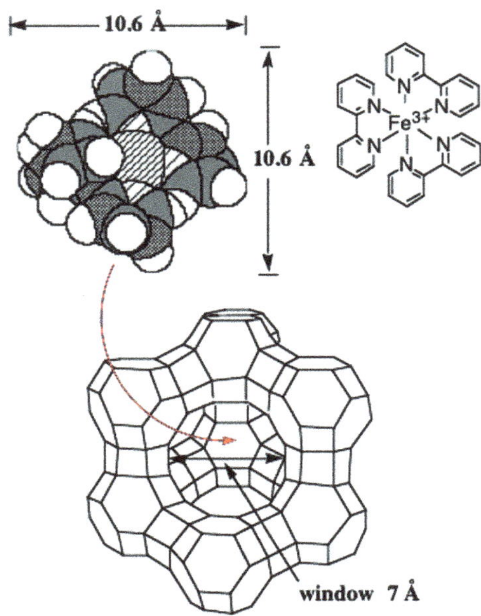

Figure 2.1 Incorporation of $[Fe(bpy)_3]^{3+}$ into the zeolite Y supercage by ship-in-bottle synthesis. Reproduced from ref. 8 with permission from American Chemical Society, Copyright 2001.

the zeolite in MeCN at 298 K is extremely slow, as shown in Figure 2.2, where the ET was far from completion in 4 h.[8]

The rate constants of ET at the averaged distance (k_{ETav}) were determined based on the statistical analysis on the time courses of electron transfer from a series of electron donors to $[Fe(bpy)_3]^{3+}$ in the zeolite assuming a normal Gaussian distribution of r. By using the β value of 1.1 $Å^{-1}$ [eqn (2.21)],[9] the average distance for ET from $Fe(C_5H_5)_2$ to $[Fe(bpy)_3]^{3+}$-zeolite Y is estimated to be 33 Å. A plot of $\log k_{ETav}$ *vs.* $-\Delta G_{ET}$ in Figure 2.2 shows parabolic dependence of $\log k_{ETav}$ on $-\Delta G_{ET}$, as expected from eqn (2.20). The k_{ETav} value increases with increasing the ET driving force to reach a maximum ($-\Delta G_{et} = \lambda$) and then decreases in order $Fe(C_5H_5)_2 > Fe(C_5Me_5)_2 > Mn(C_5Me_5)_2 > Co(C_5H_5)_2$ with increasing the ET driving force, which is regarded as the Marcus inverted region (Figure 2.3).[8] The λ value (1.0 eV) in Figure 2.3 agrees with that λ value of electron self-exchange between ferrocene and ferrocenium cation.[10]

ET reactions of the reversed direction are also examined by using $[Fe(bpy)_3]^{2+}$-zeolite Y as an electron donor instead of $[Fe(bpy)_3]^{3+}$-zeolite Y. ET from $[Fe(bpy)_3]^{2+}$-zeolite Y to $[Ru(bpy)_3]^{3+}$ and $[Ru(Me_2bpy)_3]^{3+}$ was also examined, and the k_{ETav} values also fit the

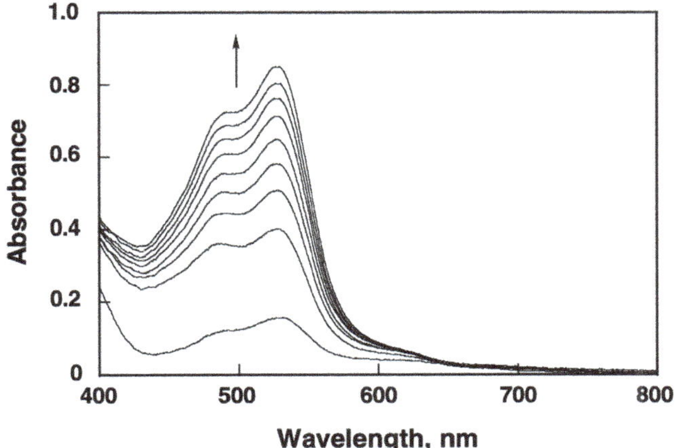

Figure 2.2 Visible absorption spectral change observed in ET from $Fe(C_5H_5)_2$ (1.0×10^{-3} M) to $[Fe(bpy)_3]^{3+}$-Y (5.4×10^{-5} mol g^{-1}) in deaerated MeCN at 298 K. Spectra were recorded at 0, 1, 5, 15, 30, 60, 90, 120 and 180 min after mixing $Fe(C_5H_5)_2$ with $[Fe(bpy)_3]^{3+}$-Y. Reproduced from ref. 8 with permission from American Chemical Society, Copyright 2001.

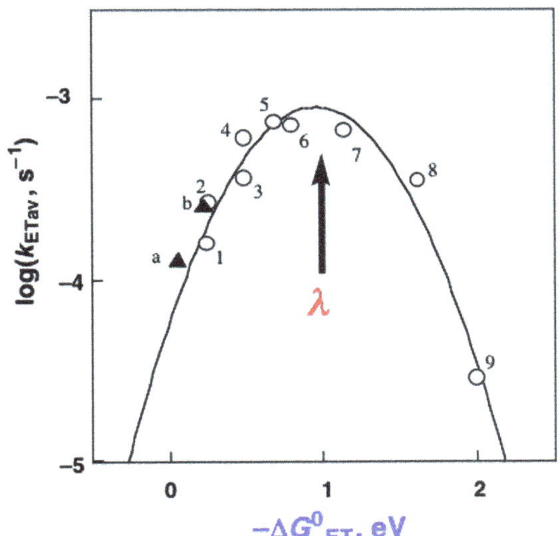

Figure 2.3 Dependence of logarithm of the rate constants of ET form electron donor to $[Fe(bpy)_3]^{3+}$-zeolite Y at the averaged distance ($\log k_{ETav}$) on ET driving force ($-\Delta G_{ET}$) in deaerated MeCN at 298 K. (○) ET from electron donors to $[Fe(bpy)_3]^{3+}$-zeolite Y (1: $Fe(C_5H_4COCH_3)_2$, 2: 10-methyl-9,10-dihydroacridine, 3: $Fe(C_5H_5)(C_5H_4COCH_3)$, 4: 1-benzyl-1,4-dihydronicotinamide, 5: $Fe(C_5H_5)_2$, 6: $Fe(C_5H_4Me)_2$, 7: $Fe(C_5Me_5)_2$, 8: $Mn(C_5Me_5)_2$, 9: $Co(C_5H_5)_2$). (▲) Electron transfer from $[Fe(bpy)_3]^{2+}$-zeolite Y to electron acceptors (a: $[Ru(Me_2bpy)_3]^{3+}$, b: $Ru[(bpy)_3]^{3+}$). Reproduced from ref. 8 with permission from American Chemical Society, Copyright 2001.

Marcus parabolic relation in Figure 2.3.[8] Thus, the existence of the Marcus inverted region in the long-range (nonadiabatic) ET has been clearly demonstrated.

The Marcus inverted region was also clearly observed for intermolecular ET reactions from fullerenes to arene radical cations (*vide infra*). Pulse radiolysis of a CH_2Cl_2 solution of arenes and fullerenes was performed to determine the rate constants of intermolecular ET from fullerenes to arene radical cations [eqn (2.22)] which were produced by the ET oxidation of arenes with $CH_2Cl_2^{\bullet+}$ that resulted from the primary ionization of CH_2Cl_2.[11]

$$\text{fullerene} + \text{arene}^{\bullet+} \rightarrow \text{fullerene}^{\bullet+} + \text{arene} \tag{2.22}$$

The driving force dependence of $\log k_{et}$ for intermolecular ET from fullerenes (C_{60}, C_{76}, and C_{78}) to a series of arene radical cations in CH_2Cl_2 at 298 K is shown in Figure 2.4(left), which shows the existence of the Marcus inverted region after levelling off by diffusion,

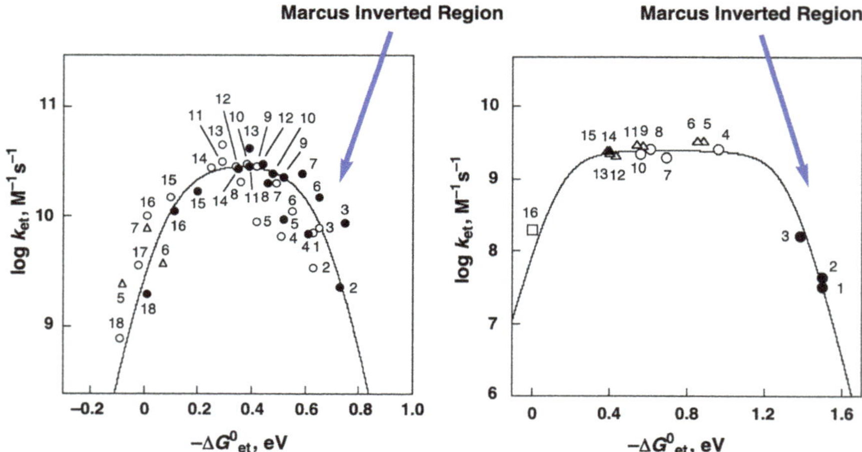

Figure 2.4 Left: Plot of $\log k_{et}$ *versus* $-\Delta G_{et}$ for intermolecular ET from C_{60} (△), C_{76} (○), and C_{78} (●) to arene radical cations in CH_2Cl_2 at 298 K. The solid line is calculated and drawn using eqn (2.16). Right: Plot of $\log k_{et}$ *vs.* $-\Delta G_{et}$ for intermolecular ET from anthracene radical anion to C_{60}, C_{70} and 1,4-^{t}Bu(PhCH$_2$)C_{60} in PhCN (○), ET from $C_{60}^{\bullet-}$ to *p*-chloranil in PhCN (△),[12] ET from NADH analogues to $^3C_{60}{}^*$ and $^3C_{70}{}^*$ in PhCN (△),[13,14] electron self-exchange between ^{t}Bu$C_{60}^{\bullet}$ and ^{t}BuC_{60}^- in PhCN/toluene (□)[15] and ET from metalloporphyrin radical anions to C_{60} (○) in propanol/toluene/acetone at 298 K. The solid line is drawn using eqn (2.16). Reproduced from ref. 11 with permission from John Wiley and Sons, © 2002 Wiley-VCH Verlag GmbH & Co. KGaA, Weinheim.

where the k_{et} value decreases with increasing the ET driving force.[11] The ET reorganization energy in CH_2Cl_2 is estimated to be as small as 0.36 eV.[11]

The driving force dependence of $\log k_{et}$ for intermolecular ET from a series of π-radical anions (metalloporphyrin radical anions and anthracene radical anions) to fullerenes (C_{60}, C_{70} and 1,4-^{t}Bu(PhCH$_2$)C_{60}) in benzonitrile (PhCN) at 298 K is also shown in Figure 2.4(right), which further confirms the existence of the Marcus inverted region after levelling off by diffusion.[11] The ET reorganization energy in PhCN is estimated to be 0.72 eV.[11] The larger λ value for the ET reduction of fullerenes in PhCN (0.72 eV) than that for the ET oxidation of fullerenes in CH_2Cl_2 (0.36 eV) may result from the larger solvent reorganization in PhCN, which is a more polar solvent than CH_2Cl_2, because the reorganization energy (λ) is composed of the bond reorganization energy and the solvent reorganization energy (*vide infra*).[11] The small overall ET reorganization energies of fullerenes have made it possible to observe unequivocally the Marcus inverted region for intermolecular ET reactions ($-\Delta G_{et} > \lambda$ in Figure 2.4).[11]

The small ET reorganization energy of fullerenes was also confirmed by determination of the electron self-exchange reaction between *tert*-butyl-C_{60} radical adduct (*t*-Bu^t$C_{60}^{\bullet}$) and the one-electron reduced species (*t*-BuC_{60}^{-}) [eqn (2.23)] (*vide infra*).[15] *t*-Bu$C_{60}^{\bullet}$ was produced by UV irradiation of a benzonitrile/benzene (1 : 7 v/v) solution of C_{60} and *tert*-butyl iodide (*t*-BuI).[16] The EPR spectrum of *t*-Bu$C_{60}^{\bullet}$ consists of 10 lines with the binomial intensity distribution resulting from nine equivalent protons of the *t*-Bu group interacting with a single unpaired electron, a(9H) = 0.17 G, as shown in the upper part of Figure 2.5. When *t*-BuC_{60}^{-}, produced by the reaction of C_{60}^{2-} with *t*-BuI,[17] was added to a benzonitrile/benzene (1 : 7 v/v) solution of C_{60} and *t*-BuI under UV irradiation, the maximum slope linewidth (ΔH_{msl}) of each EPR signal line of *t*-Bu$C_{60}^{\bullet}$ increased linearly with an increase in concentration of *t*-BuC_{60}^{-} due to electron self-exchange between *t*-Bu$C_{60}^{\bullet}$ and *t*-BuC_{60}^{-} (Figure 2.5).[15] The rate constant (k_{ex}) of the electron exchange reaction between *t*-BuC_{60}^{-} and *t*-Bu$C_{60}^{\bullet}$ [eqn (2.23)] was determined to be 1.9×10^8 M^{-1} s^{-1} at 298 K using eqn (2.24), where ΔH_{msl} and ΔH^0_{msl} are the maximum slope linewidths of the ESR signals in the presence and absence of *t*-BuC_{60}^{-}, respectively, and P_i is a statistical factor.[18,19] The corresponding λ value of the electron self-exchange reaction between *t*-BuC_{60}^{-} and *t*-Bu$C_{60}^{\bullet}$ was determined to be 0.64 eV. It should be noted that this λ value is smaller than that of the ET oxidation of C_{60} by arene radical cations in PhCN (0.72 eV).[11]

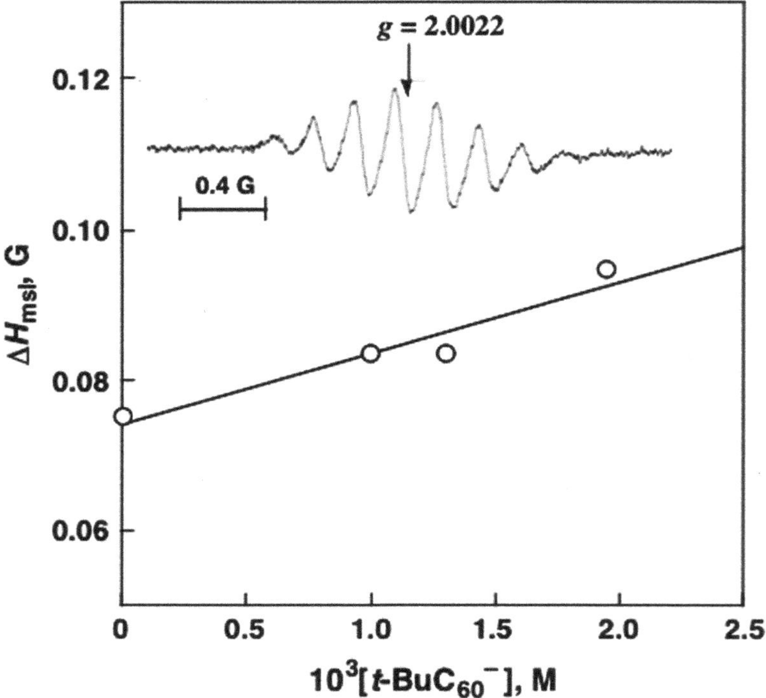

Figure 2.5 Plot of ΔH_{msl} of EPR signals due to t-BuC$_{60}$$^\bullet$ *vs.* [t-BuC$_{60}$$^-$] for EPR spectra of t-BuC$_{60}$$^\bullet$ in the absence and presence of t-BuC$_{60}$$^-$. Upper part: EPR spectrum of t-BuC$_{60}$$^\bullet$ produced by UV irradiation of a deaerated PhCN/benzene solution (1 : 7 v/v) of C_{60} and t-BuI at 298 K. Reproduced from ref. 15 with permission from American Chemical Society, Copyright 2003.

$$t\text{-BuC}_{60}{}^{-} + t\text{-BuC}_{60}{}^{\bullet} \xrightarrow{k_{ex}} t\text{-BuC}_{60}{}^{\bullet} + t\text{-BuC}_{60}{}^{-} \tag{2.23}$$

$$k_{ex} = (1.52 \times 10^{7})(\Delta H_{msl} - \Delta H^{0}_{msl})/\{(1 - P_{i})[t\text{–BuC}_{60}{}^{-}]\} \tag{2.24}$$

The nonexistence and existence of the Marcus inverted region have been studied in photoinduced ET from electron donors (D) to the singlet excited state of a flavin analogue (10-methylisoalloxazine: ^{1}MeFl*) and the back ET reactions, as shown in Scheme 2.6, where k_{diff} and k_{-diff} are diffusion and dissociation rate constants in the encounter complex (DMeFl*), respectively.[20] The rate constant of forward intramolecular photoinduced ET in the encounter complex,

D + (MeFl*) $\underset{k_{-diff}}{\overset{k_{diff}}{\rightleftharpoons}}$ (D MeFl*)

$\downarrow k_{ET}$

($D^{\bullet +}$ $MeFl^{\bullet -}$) $\underset{k_{-diff}}{\overset{k_{diff}}{\rightleftharpoons}}$ $D^{\bullet +}$ + $MeFl^{\bullet -}$

$\downarrow k_{BET}$

D + MeFl

Scheme 2.6 Photoinduced ET from electron donors (D) to the singlet excited state of MeFl (1MeFl*) and the subsequent reactions. Reproduced from ref. 20 with permission from John Wiley and Sons, Copyright 2010 Wiley-VCH Verlag GmbH & Co. KGaA, Weinheim.

(DMeFl*) to ($D^{\bullet +}MeFl^{\bullet -}$), is denoted as k_{ET}, whereas the rate constant of the forward intermolecular photoinduced ET, from (D + MeFl*) to ($D^{\bullet +}MeFl^{\bullet -}$), is denoted as k_{et}.[20] The rate constant of the back ET (BET) to the ground state by intramolecular and intermolecular BET are denoted as k_{BET} and k_{bet}, respectively.[20]

The driving force dependence of $\log k_{et}$ of ET from D to ^{1}MeFl* is shown in Figure 2.6, where the k_{et} values increase with increasing the ET driving force to reach the k_{diff} value, agreeing well with the solid line drawn using eqn (2.16) with use of the λ value of 0.59 eV in PhCN.[20] This λ value is smaller than that of 0.66 eV for $C_{60}/C_{60}^{\bullet -}$ in PhCN[11] and 0.88 eV for 9-phenyl-10methylacridinium ion ($AcrPh^+$)/ $AcrPh^{\bullet}$ in MeCN).[21] The calculated driving force dependence of k_{et} based on eqn (2.16) predicts a decrease in the k_{et} value from a diffusion-limited value with increasing the ET driving force ($-\Delta G_{et} > 1.0$ eV), provided that the λ value remains the same (0.59 eV). However, the observed k_{et} values remains the diffusion-limited value up to $-\Delta G_{et} = 1.74$ eV (Figure 2.6).[20] The absence of the Marcus inverted region may result from the formation of the excited state products.[20] Photoinduced ET from D (**14–17**) to MeFl with large ET driving forces results in formation of the doublet excited state of $D^{\bullet +}*$.[20] In such a case, the actual ET driving force to produce $D^{\bullet +}*$ is not located in the Marcus inverted region but, shifted to the Marcus normal region.[20] The driving forces [$-\Delta G_{et}(S)$ and $-\Delta G_{et}(T)$] of photoinduced ET from D (**14–17**) to the singlet and triplet excited states of MeFl to produce

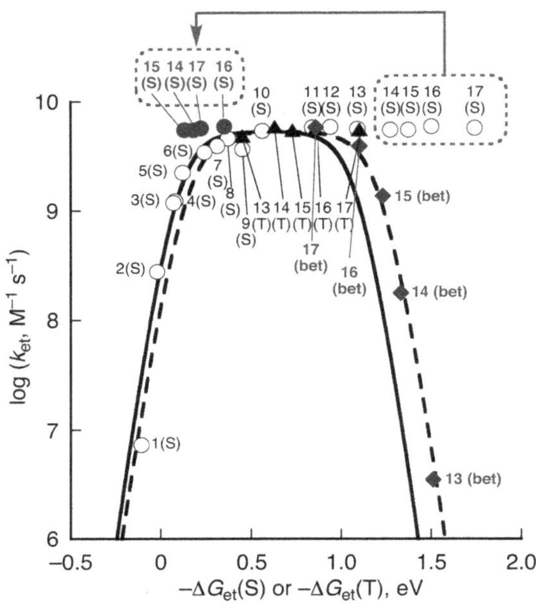

Figure 2.6 Driving force dependence of log k_{et} for intermolecular photoinduced ET from electron donors to ^{1}MeFl* ($-\Delta G_{et}$(S): ○ and ●) and ^{3}MeFl* ($-\Delta G_{et}$(T): s), and log k_{bet} for intermolecular back ET ($-\Delta G_{bet}$: ◆) in PhCN at 298 K. Open circle symbols (○): highly exergonic ET reactions to produce ground-state products (radical cations of electron donors and $MeFl^{\bullet-}$); black circle symbols (●): the much less exergonic ET reactions to produce the excited states of radical cations of electron donors and $MeFl^{\bullet-}$. The solid and broken lines are calculated and drawn based on eqn (2.16) with use of $\lambda = 0.59$ and 0.68 eV, respectively. Numbers refer to the electron donors. 1: 1,2,3-trimethylbenzene, 2: 1,2,4-trimethylbenzene, 3: 1,2,3,4-tetramethylbenzene, 4: 1,2,3,5-tetramethylbenzene, 5: 1,2,4,5-tetramethylbenzene, 6: hexamethylbenzene, 7: 4-methoxybenzene, 8: *trans*-stilbene, 9: 1,2,3-trimethoxybenzene, 10: 1,4-dimethoxybenezene, 11: 1,2,4-trimethoxybenzene, 12: triphenylamine, 13: *N*,*N*-dimethylaniline, 14: phenothiazine, 15: bis(ethylenedithiol)tetrathiafulvalene, 16: *N*,*N*,*N*′,*N*′-tetramethylbenzidine, 17: *N*,*N*,*N*′,*N*′-tetramethylphenylendiamine. Reproduced from ref. 20 with permission from John Wiley and Sons, Copyright © 2010 Wiley-VCH Verlag GmbH & Co. KGaA, Weinheim.

$D^{\bullet+*}$ and $MeFl^{\bullet-}$ are determined by subtracting the energy of the low-lying excited state of $D^{\bullet+}$ ($^2E^*$) according to eqn (2.25) and (2.26), where $^1E^*$ and $^3E^*$ are the singlet and triplet excited state energies, respectively.[20] The first low-lying excited states of phenothiazine and *N*,*N*,*N*′,*N*′-tetramethylphenylendiamine radical cations were reported to be 1.09 eV and 1.52 eV, respectively.[22,23] Only the vertical $^2E^*$ values are available because of the lack of the fluorescence data of $D^{\bullet+*}$. The k_{et} values of photoinduced ET from D (**14–17**: closed circles) to ^{1}MeFl* to produce the excited state $D^{\bullet+*}$ and $MeFl^{\bullet-}$ agree reasonably well with the broken line in Figure 2.6, predicted with use of eqn (2.16), because the actual driving forces of photoinduced ET from D (**14–17**) to MeFl to produce $D^{\bullet+*}$ and $MeFl^{\bullet-}$ may be larger than the estimated values using the vertical $^2E^*$ values.[20]

$$-\Delta G_{\mathrm{et}}(\mathrm{S}) = {}^1E^* - e(E_{\mathrm{ox}} - E_{\mathrm{red}}) - {}^2E^* \tag{2.25}$$

$$-\Delta G_{\mathrm{et}}(\mathrm{T}) = {}^3E^* - e(E_{\mathrm{ox}} - E_{\mathrm{red}}) \tag{2.26}$$

In contrast to the case of photoinduced ET reactions of MeFl, the driving force dependence of the logarithm of the rate constant of intermolecular back ET (BET) from $MeFl^{\bullet-}$ to $D^{\bullet+}$ ($\log k_{\mathrm{bet}}$) is shown in Figure 2.6 (nos. 13–17), which clearly reveals the existence of the Marcus inverted region. The data are well fitted by the Marcus equation for intermolecular outer-sphere BET reactions [eqn (2.16)] with use of the reorganization energy of BET ($\lambda = 0.68$ eV) in PhCN.[20] The λ value of the BET reactions is larger than that (0.59 eV) of the photoinduced ET reactions in the Marcus normal region, because the BET reactions occur at the longer distance as compared with the photoinduced ET reactions in the Marcus normal region and the solvent reorganization energy of BET (λ_{s}) is expected to increase with an increase in the distance (R_{DA}) between D and A as given by eqn (2.27), where e is the electric charge, ε_0 is the dielectric constant in a vacuum, r_{D} and r_{A} are the ionic radii of D and A, and $\varepsilon_{\mathrm{op}}$ and ε_{s} are the optical and static dielectric constants of the solvent, respectively.[1]

$$\lambda = \frac{e^2}{4\pi\varepsilon_0}\left(\frac{1}{2r_{\mathrm{D}}} + \frac{1}{2r_{\mathrm{A}}} - \frac{1}{R_{\mathrm{DA}}}\right)\left(\frac{1}{\varepsilon_{\mathrm{op}}} - \frac{1}{\varepsilon_{\mathrm{s}}}\right) \tag{2.27}$$

The driving force dependence of intramolecular photoinduced charge separation (CS) and charge recombination (CR) of an electron donor–acceptor dyad (D–A) (Figure 2.7a) is shown in Figure 2.7b.[4] When the magnitude of the ET driving force becomes the same as the reorganization energy ($-\Delta G_{\mathrm{ET}} = \lambda$), the ET rate reaches a maximum. The maximum value is determined by the magnitude of electronic coupling (V) between the donor and acceptor moiety [eqn (2.20)].[4] When the reorganization energy becomes smaller, the rate constant of CS becomes larger, whereas the rate constant of CR becomes smaller (Figure 2.7b) in the Marcus inverted region ($-\Delta G_{\mathrm{ET}} > \lambda$).[4] In such a case, the larger the driving force, the smaller becomes the rate constant of CR when the CS lifetime becomes longer.[4] Thus, the

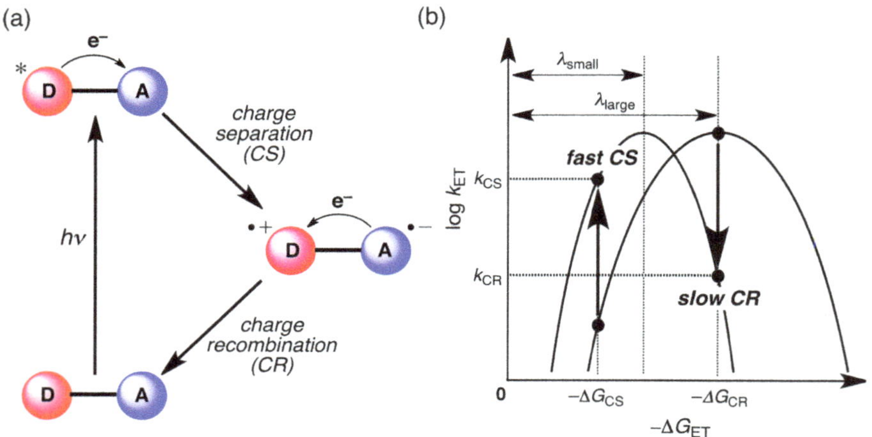

Figure 2.7 (a) Energy diagram of photoinduced charge separation (CS) and charge recombination (CR) occurring in an electron donor–acceptor (D–A) linked dyad. (b) Driving force dependence of $\log k_{ET}$ of CS and CR with different λ values.

magnitude of the reorganization energy is the key parameter to control the CS and CR processes.

References

1. R. A. Marcus, *Annu. Rev. Phys. Chem.*, 1964, **15**, 155–296.
2. R. A. Marcus, *Angew. Chem., Int. Ed. Engl.*, 1993, **32**, 1111–1121.
3. R. A. Marcus and N. Sutin, *Biochim. Biophys. Acta, Rev. Bioenerg.*, 1985, **811**, 265–322.
4. S. Fukuzumi, *Electron Transfer: Mechanisms and Applications*, Wiley-VCH, 2020.
5. J. R. Miller, L. T. Calcaterra and G. L. Closs, *J. Am. Chem. Soc.*, 1984, **106**, 3047–3049.
6. G. L. Closs and J. R. Miller, *Science*, 1988, **240**, 440–447.
7. I. R. Gould and S. Farid, *Acc. Chem. Res.*, 1996, **29**, 522–528.
8. S. Fukuzumi, Y. Yoshida, T. Urano, T. Suenobu and H. Imahori, *J. Am. Chem. Soc.*, 2001, **123**, 11331–11332.
9. W. J. Albery, P. N. Bartlett, C. P. Wilde and J. R. Darwent, *J. Am. Chem. Soc.*, 1985, **107**, 1854–1858.
10. E. S. Yang, M.-S. Chan and A. C. Wahl, *J. Phys. Chem.*, 1980, **27**, 3094–3099.
11. S. Fukuzumi, K. Ohkubo, H. Imahori and D. M. Guldi, *Chem. – Eur. J.*, 2003, **9**, 1585–1593.
12. J. W. Arbogast, C. S. Foote and M. Kao, *J. Am. Chem. Soc.*, 1992, **114**, 2277–2279.
13. S. Fukuzumi, T. Suenobu, M. Patz, T. Hirasaka, S. Itoh, M. Fujitsuka and O. Ito, *J. Am. Chem. Soc.*, 1998, **120**, 8060–8068.
14. S. Fukuzumi, T. Suenobu, T. Hirasaka, N. Sakurada, R. Arakawa, M. Fujitsuka and O. Ito, *J. Phys. Chem. A*, 1999, **103**, 5935–5941.
15. S. Fukuzumi, I. Nakanishi, T. Suenobu and K. M. Kadish, *J. Am. Chem. Soc.*, 1999, **121**, 3468–3474.
16. J. R. Morton, F. Negri and K. F. Preston, *Acc. Chem. Res.*, 1998, **31**, 63–69.

17. S. Fukuzumi, T. Suenobu, T. Hirasaka, R. Arakawa and K. M. Kadish, *J. Am. Chem. Soc.*, 1998, **120**, 9220–9227.
18. R. Chang, *J. Chem. Educ.*, 1970, **47**, 563–568.
19. K. S. Cheng and N. Hirota, in *Investigation of Rates and Mechanisms of Reactions*, ed. G. E. Hammes, Wiley-Interscience, New York, 1974, vol. VI, p. 565.
20. M. Murakami, K. Ohkubo and S. Fukuzumi, *Chem. – Eur. J.*, 2010, **16**, 7820–7832.
21. S. Fukuzumi, K. Ohkubo, T. Suenobu, K. Kato, M. Fujitsuka and O. Ito, *J. Am. Chem. Soc.*, 2001, **123**, 8459–8467.
22. L. N. Domelsmith, L. L. Munchausen and K. N. Houk, *J. Am. Chem. Soc.*, 1977, **99**, 6506–6514.
23. R. Egdell, J. C. Green and C. N. R. Rao, *Chem. Phys. Lett.*, 1975, **33**, 600–607.

3 Photosynthetic Reaction Centre Models

3.1 Photosynthetic Reaction Centre

The photoinduced electron transfer (ET) processes in a membrane-bound protein in the photosynthetic reaction centre (PRC), which contains a number of cofactors, including four bacteriochlorophylls (BChl), are shown in Figure 3.1 with the structure of the PRC.[1–6] The photosynthetic reaction centre is surrounded by light-harvesting units. After efficient energy transfer in the light-harvesting centre, energy transfer to the special pair [$(BChl)_2$] results in generation of a singlet excited state. Then, electron transfer from the singlet excited state to bacteriopheophytin (BPhe) yields the first charge-separated (CS) state [$(BChl)_2^{\bullet+}BPhe^{\bullet-}$] with a time constant of 3 ps.[7] Without the other components, the two ubiquinones (Q_A and Q_B), the CS state is converted back to the ground state by back electron transfer (BET) with a 10 ns lifetime.[7] ET from the initial CS state to Q_A proceeds with a much faster rate than the BET rate to produce the second CS state ($(BChl)_2^{\bullet+}Q_A^{\bullet-}$) despite the ET driving force being smaller than the BET driving force, because the BET is located deep in the Marcus inverted region (*vide supra*).[7] Then, a further CS process, that is ET from $Q_A^{\bullet-}$ to $Q_B^{\bullet-}$, proceeds with a much faster rate than the BET to afford the final CS state [$(BChl)_2^{\bullet+}Q_A^{\bullet-}$] with a nearly 100% quantum yield and a lifetime of one second, which is long enough for further chemical reactions.[8] The long lived CS state results from the

RSC Foundations No. 1
Artificial Photosynthesis
By Shunichi Fukuzumi

Published by the Royal Society of Chemistry, www.rsc.org

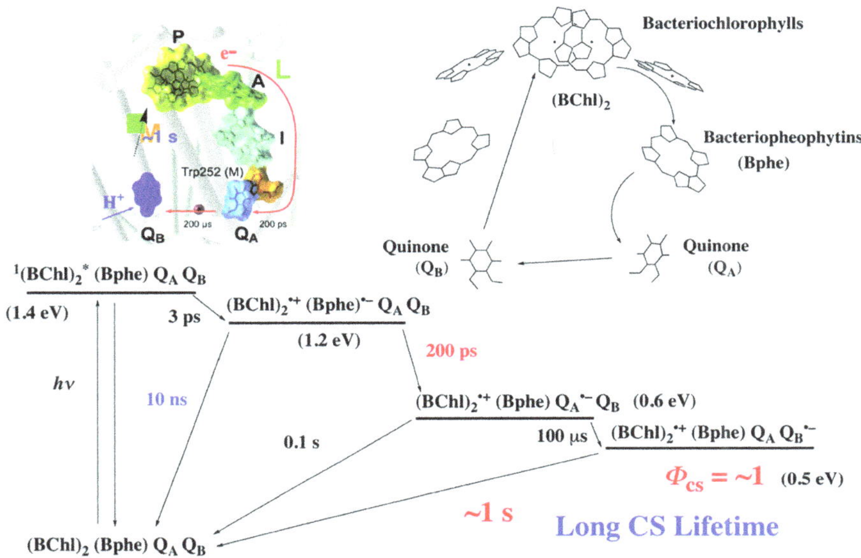

Figure 3.1 Structure and energy diagram of the PRC of purple bacteria.

long distance in the back electron transfer from $Q_A^{•-}$ to $(BChl)_2^{•+}$ [eqn (2.20) and (2.21)].

3.2 Electron Donor–Acceptor Linked by Covalent Bonds

3.2.1 Porphyrins

The first candidates for components of photosynthetic reaction centre (PRC) models are chlorophylls and their analogues (chlorins and porphyrins), which are involved in the PRC and other biological ET systems. Chlorophylls, chlorins and porphyrins contain a highly conjugated two-dimensional π-system, which is suitable for efficient ET reactions as electron donors and acceptors because the ET reduction and ET oxidation of the two-dimensional π-system may be associated with a very small structural and solvation change to afford small λ values of the ET reactions.[9,10] Rich and extensive visible absorption features of porphyrinoid systems are also suitable for an efficient use of the solar spectrum.[9,10] On the other hand, fullerenes such as C_{60}, which contain a highly conjugated three-dimensional π-system, are also suitable for efficient ET reactions as electron acceptors because the uptake of electrons also results in minimal structural and solvation changes to afford small λ values of ET reactions.[11,12]

The first successful example of an electron donor–acceptor dyad linked with a covalent bond as a model of the initial CS state in the PRC is composed of zinc chlorin and C_{60} (ZnCh–C_{60} dyad), as shown in Figure 3.2, where the laser excitation of a PhCN solution of ZnCh–C_{60} at 355 nm resulted in intramolecular ET from the singlet excited state of ZnCh to C_{60} to produce the CS state ($ZnCh^{\bullet +}$–$C_{60}{}^{\bullet -}$).[13] The absorption maxima at 790 and 1000 nm are assigned due to $ZnCh^{\bullet +}$ and $C_{60}{}^{\bullet -}$, respectively.[13] The absorbance is due to the CS state decayed *via* BET [charge recombination (CR)] to the ground-state rather than to the triplet excited state ($^3ZnCh^*$ or $^3C_{60}{}^*$).[13] The CR rate was determined from the disappearance of the absorption band at 790 nm due to $ZnCh^{\bullet +}$ in $ZnCh^{\bullet +}$–$C_{60}{}^{\bullet -}$, which obeyed first-order kinetics. The CR rate constant was determined to be 9.1×10^3 s^{-1}, which corresponds to the CS lifetime of 110 μs.[13] This CS lifetime is

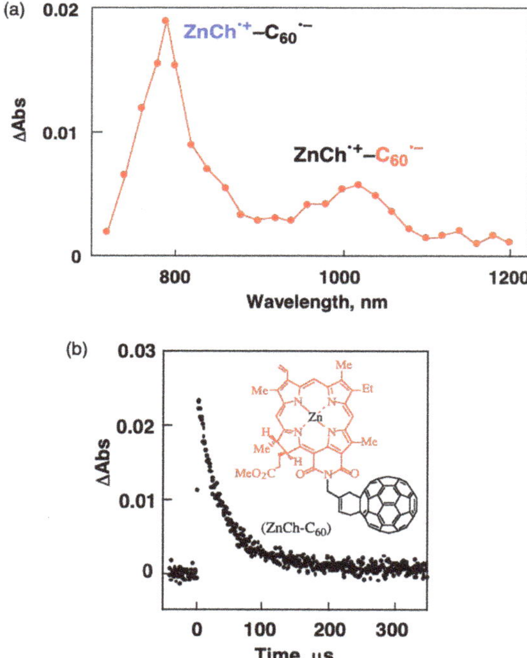

Figure 3.2 (a) Transient absorption spectrum observed at 0.1 μs after laser excitation at 355 nm of a deaerated PhCN solution of ZnCh–C_{60} (1.0×10^{-4} M) at 298 K. (b) The decay time profile of absorbance at 790 nm due to the $ZnCh^{\bullet +}$ moiety of $ZnCh^{\bullet +}$–$C_{60}{}^{\bullet -}$. Reproduced from ref. 13 with permission from American Chemical Society, Copyright 2001.

much longer than the initial CS lifetime without other electron acceptors in the PRS (10 ns, in Figure 3.1).[13]

A bacteriochlorin-C_{60} dyad ($H_2BCh–C_{60}$ (**1**)) and $ZnCh–C_{60}$ (**2**) dyad (Figure 3.3) were synthesized *via* a Diels–Alder reaction of the chlorin and porphyrin with a C_{60} derivative, and the structures of **1** and **2** are shown together with the reference compounds (**3** and **4**).[13] The dyads (**1**, **2**, **5–8**) have the same short spacer, where the edge-to-edge distance (R_{ee}) is 5.9 Å, whereas the R_{ee} value of ZnPCONH–C_{60} (**9**) is much longer (11.9 Å).[15] The driving force ($-\Delta G_{ET}$ or $-\Delta G_{BET}$) dependence of log k_{ET} and k_{BET} of the dyads (**1,2**, **5–9**) is shown in Figure 3.4, where the best fit line based on the Marcus equation [eqn (2.20)] affords

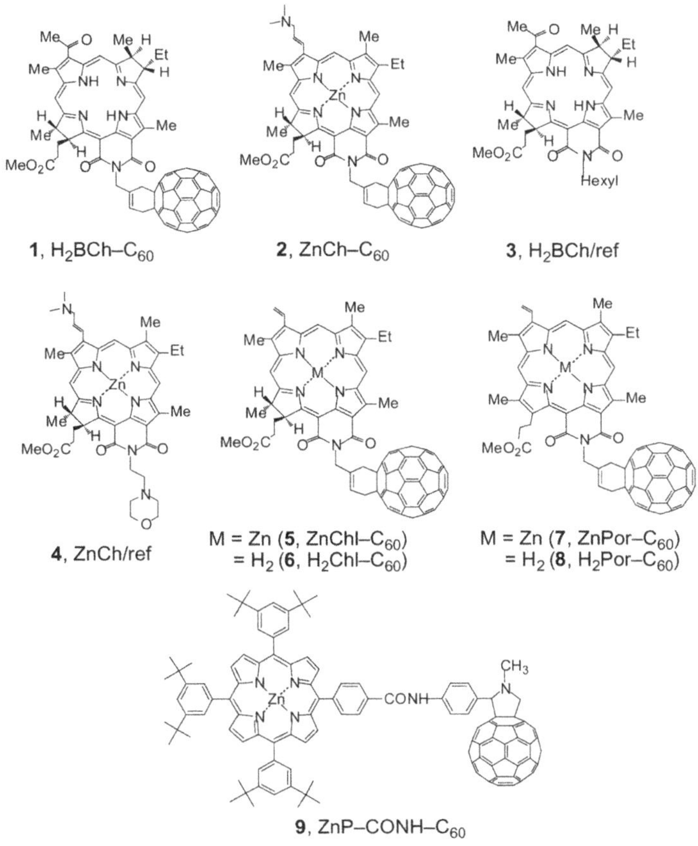

1, $H_2BCh–C_{60}$ **2**, $ZnCh–C_{60}$ **3**, H_2BCh/ref

4, ZnCh/ref M = Zn (**5**, ZnChl–C_{60}) = H_2 (**6**, $H_2Chl–C_{60}$) M = Zn (**7**, ZnPor–C_{60}) = H_2 (**8**, $H_2Por–C_{60}$)

9, ZnP–CONH–C_{60}

Figure 3.3 Structure of free-base bacteriochlorin- and zinc chlorin-fullerene dyads. The structures of 1 and 2 are shown together with the reference compounds (3 and 4). The dyads (1, 2, 5–8) have the same short spacer, where the edge-to-edge distance (R_{ee}) is 5.9 Å, whereas the R_{ee} value of ZnPCONH-C_{60} (9) is much longer (11.9 Å).[15]

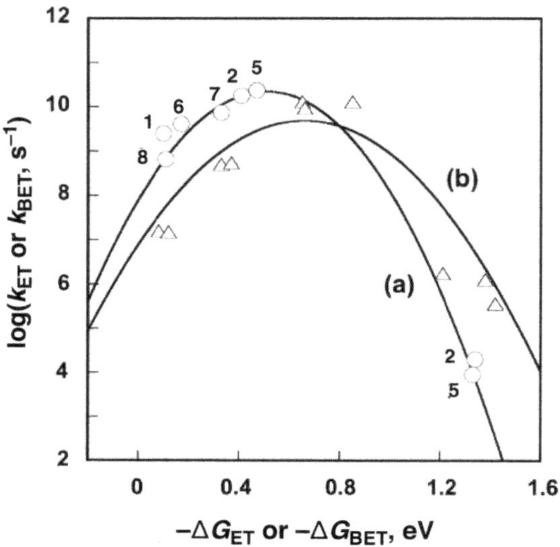

Figure 3.4 ET or BET driving force ($-\Delta G_{ET}$ or $-\Delta G_{BET}$) dependence of the logarithm of the rate constants of intramolecular ET or BET ($\log k_{ET}$ or $\log k_{BET}$) in (a) C_{60}-linked dyads (1,2, 5–8) (○) and (b) ZnP-CONH-C_{60} dyads (△) in PhCN.[15] The solid lines are drawn using eqn (2.20) with the values of (a) $\lambda = 0.51$ eV and $V = 7.8$ cm^{-1}, and (b) $\lambda = 0.66$ eV and $V = 3.9$ cm^{-1}. Numbers refer to the compounds in Figure 3.3.

$\lambda = 0.51$ eV and $V = 7.8$ cm^{-1}.[13] On the other hand, the best fit line for the zinc porphyrin linked fullerene dyads with longer amide linkage (ZnP–CONH–C_{60}, **9**)[15] affords $\lambda = 0.66$ eV and $V = 3.9$ cm^{-1}.[13] In both cases, the CS processes are located in the Marcus normal region ($-\Delta G_{ET} < \lambda$), whereas the BET processes are located in the Marcus inverted region ($-\Delta G_{ET} > \lambda$). The solvent reorganization energy is expected to decrease with decreasing the donor–acceptor distance [eqn (2.27)].[13] Thus, the smaller λ value (0.51 eV) of the dyads with the shorter spacers leads to the smaller k_{BET} value as compared to the dyads with the longer spacers (Figure 3.4).[13] On the other hand, the V value of the shorter spacer (7.8 cm^{-1}) is larger when the k_{ET} value is larger as compared to the longer spacer (3.9 cm^{-1}).[13]

In the case of H_2BCh, the energy of the CS state is higher than that of the triplet excited state of bacteriochlorin ($^3H_2BCh^*$), as shown in Scheme 3.1a, where BET from $C_{60}^{\bullet -}$ to $H_2BCh^{\bullet +}$ produces $^3H_2BCh^*$ rather than the ground state (H_2BCh) because the BET to produce $^3H_2BCh^*$ is located in the Marcus normal region, whereas the BET to the ground state is located deep in the Marcus inverted region. In such a case, the BET to the ground state is too slow to compete with the BET

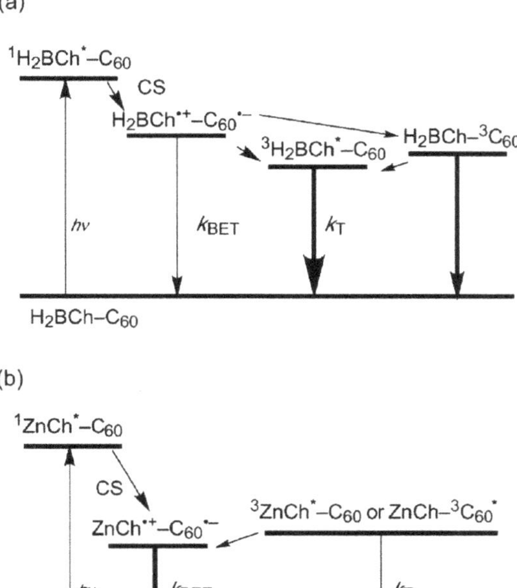

Scheme 3.1 Energy diagrams of CS and BET (CR) of (a) H_2Bh-C_{60} and (b) $ZnCh-C_{60}$: the importance of the triplet excited state energy in comparison with the CS state energy to obtain the long-lived CS state. Reproduced from ref. 13 with permission from American Chemical Society, Copyright 2001.

to $^3H_2BCh^*$.[13] In contrast to the case of H_2BCh, the energy of the triplet excited state of $^3ZnCh^*$ is higher than that of the CS state. In such a case, the BET proceeds slowly in the Marcus inverted region to produce the ground state (no triplet excited state is produced), as shown in Scheme 3.1b.[13] Thus, it is very important to choose electron donor and acceptor moieties, which have small ET reorganization energy and to link them with a short distance to obtain the long-lived CS state.

As expected from the Marcus plots in Figure 3.4, photoexcitation of a zinc chlorin–fullerene dyad [(Zn(Ch)–C_{60}] with a very short linkage (Figure 3.5) resulted in formation of the ultra-long-lived CS state and the CS lifetime is highly temperature dependent, as shown in the Eyring plot, where the activation enthalpy was determined from the slope to be 5.4 kcal mol^{-1}.[16] Such a large activation enthalpy indicates that the BET(CR) is located deeply in the Marcus inverted region.[16] The CS lifetime was determined as long as 120 s in frozen PhCN at −150 °C.[16]

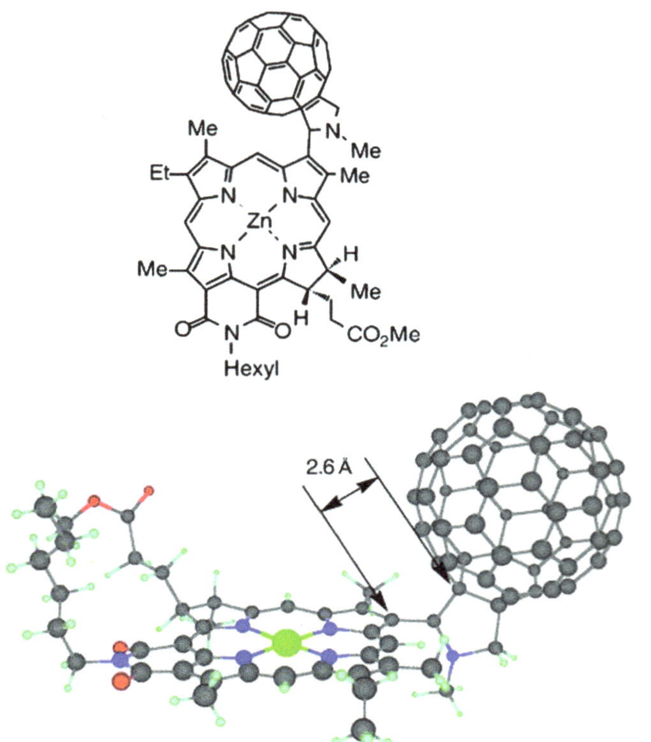

Figure 3.5 Structure of the Zn(Ch)–C_{60} dyad with short linkage and optimized structure. Reproduced from ref. 16 with permission from John Wiley and Sons, © 2002 Wiley-VCH Verlag GmbH & Co. KGaA, Weinheim.

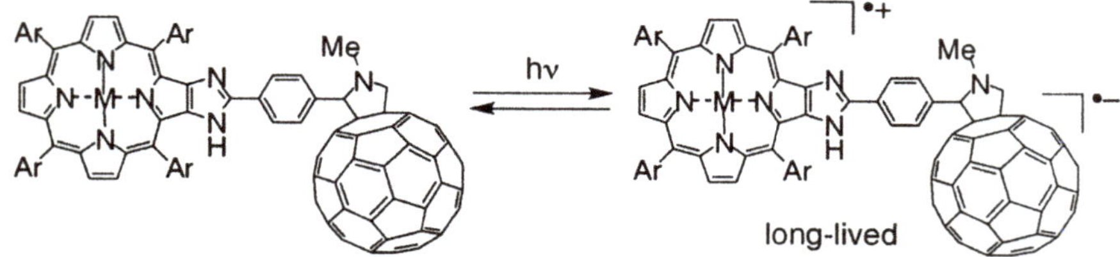

Figure 3.6 Photoinduced CS and CR in ZnImP–C_{60}. Reproduced from ref. 17 with permission from American Chemical Society, Copyright 2003.

Photoexcitation of a zinc imidazoporphyrin–fullerene dyad with a short linkage (ZnImP–C_{60}) also resulted in formation of the long-lived CS state with a lifetime of 310 μs in PhCN at 278 K (Figure 3.6).[17] The energy of the CS state [($ZnImP^{\bullet +}$–$C_{60}{}^{\bullet -}$) = 1.34 eV] is lower than that of the triplet excited states of C_{60} ($^3C_{60}{}^* = 1.50$ eV)[18] and ZnImP

(1.36 eV).[17] The rate constant of ET (k_{ET}) from the ^{1}ZnImP* moiety to the C_{60} moiety in ZnImP–C_{60} was determined to be 1.6×10^{10} s^{-1}.[17] In contrast to this, photoexcitation of the corresponding free base porphyrin dyad (H_2ImP–C_{60}) resulted in no observation of the long-lived CS state.[17] In this case, only the triplet–triplet (T–T) absorption spectrum due to $^3H_2$ImP*–C_{60} was detected, because the CS state energy (1.54 eV) is higher than the triplet excited state energy of H_2ImP (1.37 eV).[17] BET from the $C_{60}^{\bullet -}$ moiety to the $H_2ImP^{\bullet +}$ moiety produces the triplet excited state ($^3H_2$ImP*–C_{60}) rather than the ground state (H_2ImP–C_{60}), as is the case in Scheme 3.1.[17]

As described above, the solvent reorganization of ET [eqn (2.27)] plays a very important role in controlling the ET rate constant. By using the short linkage between the D–A dyad, the solvent reorganization of BET becomes smaller to attain the long CS lifetime. The solvent reorganization energy is expected to be smaller as the polarity of solvent decreases. In a nonpolar solvent, the CS energy may be too high because of much less stabilization of the CS state by solvation. In the case of a charge-shift type of ET, however, the driving force of ET from an electron donor (D) to an acceptor cation (A^+) may be insensitive to the solvent polarity, because the solvation energy due to A^+ before the ET may be cancelled largely by the solvation energy due to $D^{\bullet +}$ after the ET. This idea worked out well for the zinc porphyrin–gold porphyrin cation dyad [(ZnPQ–AuPQ$^+$(PF_6^-)] in Figure 3.7.[19] The use of a PQ ligand by annelation of qinoxaline to the porphyrin ligand resulted in a lowering of the CS state energy, which is smaller than that of the triplet excited states even in nonpolar solvents.[19] The CS state energy in toluene is determined as 1.11 eV from the one-electron oxidation potential of the ZnPQ moiety (E_{ox} *vs.* SCE = 0.81 V) and the one-electron reduction potential of the AuPQ$^+$ moiety (E_{red} *vs.*

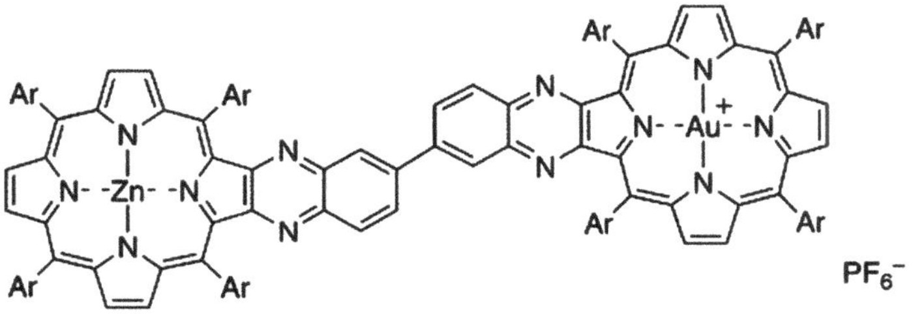

Figure 3.7 Structure of a zinc porphyrin-gold porphyrin cation dyad [(ZnPQ–AuPQ$^+$(PF_6^-)] (Ar = 3,5-^{t}Bu$_2$C$_6$H$_3$). Reproduced from ref. 19 with permission from American Chemical Society, Copyright 2003.

SCE = -0.30 V) in ZnPQ–AuPQ$^+$, which is lower than those of the triplet excited states of ZnPQ (1.32 eV) and AuPQ$^+$ (1.64 eV).[19] The CS state energy in toluene is even smaller than the value (1.21 eV) obtained from the one-electron redox potentials in PhCN.[19] The transient absorption bands observed at 650–800 nm are assigned to the charge-shifted state (ZnPQ$^{\bullet+}$–AuPQ) by comparison with the thin-layer UV-visible spectra of ZnPQ$^{\bullet+}$ and AuPQ, which were produced by the electrochemical oxidation of ZnPQ and reduction of AuPQ$^+$, respectively.[19] The lifetimes of the charge-shifted state in toluene and cyclohexane were determined to be 9.1 μs and 10 μs, respectively.[19] In contrast to the long-lived charge-shifted state in nonpolar solvents, no charge-shifted state was seen in PhCN when only the triplet–triplet absorption due to the ^{3}ZnPQ* was observed.[19] The site of the ET reduction of AuPQ$^+$ was confirmed to be the Au metal rather than the porphyrin ligand.[19]

Long-lived CS states can also be obtained with use of cup-shaped nanocarbons (CNCs) with controlled size, which were functionalized with a large number of porphyrin molecules (*vide infra*).[20] The cup-shaped nanocarbons are first functionalized with aniline as the precursor for the further functionalization with porphyrins after the reduction by naphthalene radical anions to separate each stacked nanocarbons (Figure 3.8a).[20,21] The cup-shaped nanocarbons functionalized by aniline react with the porphyrin derivatives to obtain the nanohybrids with free base porphyrins, CNC–$(H_2P)_n$ (Figure 3.8a).[20] The structure of CNC–$(H_2P)_n$ is shown by the TEM image in Figure 3.8b, which reveals cup-shaped nanocarbon with a hollow core with well-controlled diameter (*ca.* 50 nm) and size (*ca.* 100 nm).[20] The weight % of porphyrins attached to the cup-shaped nanocarbons indicates that one porphyrin molecule is attached to one out of 640 carbon atoms of the nanocup framework for CNC–$(H_2P)_n$.[20] Thus, attachment of a number of porphyrin molecules on the CNC resulted in no significant change in the π-framework of the CNC.[20] The covalent functionalization of the CNC–$(H_2P)_n$ nanohybrid was indicated by an intensity increase of the Raman signal at 1353 cm^{-1} (D band) in the functionalized CNC as compared with the pristine CSCNTs indicates,[20] because the D band has been used for indicating the process of functionalization which transforms the sp^2 carbons to sp^3 carbons.[20] The UV-vis absorption spectrum of CNC–$(H_2P)_n$ agreed with that of superposition of ref–H_2P [tetrakis(*N*-octadecyl-4-aminocarboxyphenyl)porphyrin] and cup-shaped nanocarbons, indicating that there is no significant interaction between attached porphyrins and CSCNTs in the ground states.[20]

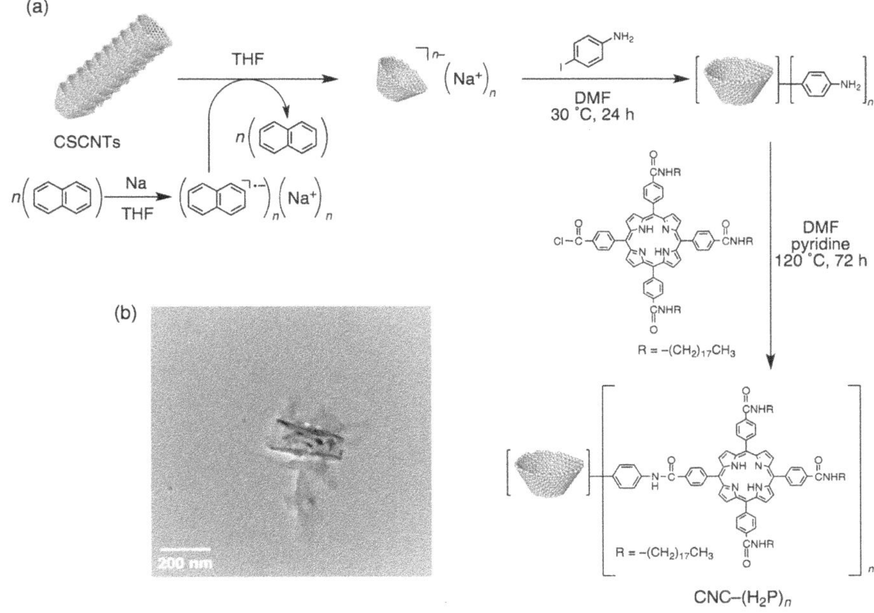

Figure 3.8 (a) Synthetic scheme of CNC-(H_2P)$_n$. (b) TEM image of CNC-(H_2P)$_n$. Reproduced from ref. 20 with permission from John Wiley and Sons, Copyright © 2009 Wiley-VCH Verlag GmbH & Co. KGaA, Weinheim.

The fluorescence lifetime of CNC–(H_2P)$_n$ (3.0 ± 0.1 ns) became much shorter than that of ref–H_2P (14.1 ± 0.1 ns).[20] The fluorescence emission at 650 nm was also quenched in CNC–(H_2P)$_n$.[20] The shortened fluorescence lifetime of CNC–(H_2P)$_n$ and the fluorescence quenching as compared to the ref–H_2P indicate that photoinduced ET proceeds from the singlet excited state of H_2P ($^1H_2P^*$) to CNC in CNC–(H_2P)$_n$ to afford the CS state (CNC$^{\bullet -}$–(H_2P)$_n$$^{\bullet +}$).[20] This was confirmed by nanosecond laser induced transient absorption measurements, where the broad transient absorption bands in the visible and near infrared (NIR) region were assigned due to $H_2P^{\bullet +}$ in CNC$^{\bullet -}$–(H_2P)$_n$$^{\bullet +}$, which was quite different form the triplet–triplet (T–T) absorption of ref–H_2P.[20] The CS state was also detected by EPR under photoirradiation of CNC–(H_2P)$_n$ in frozen *N*,*N*-dimethylformamide (DMF) at 153 K.[20] The observed isotropic EPR signal at $g = 2.0044$ agrees with that of ref–$H_2P^{\bullet +}$ produced by the ET oxidation with $[Ru(bpy)_3]^{3+}$ (bpy = 2,2′-bipyridine) in deaerated $CHCl_3$.[20] The EPR signal due to the reduced carbon-based nanomaterials (CNC$^{\bullet -}$) could not be detected probably due to significant delocalization of electron and fast relaxation time in the CNC.[20]

The decay of the CS state of CNC–$(H_2P)_n$ obeyed first-order kinetics and the first-order plots with different initial CS concentrations of the CS state afforded linear correlations with the same slope.[20] This indicates that the decay of the CS state results from intramolecular BET from $CNC^{\bullet-}$ to $H_2P^{\bullet+}$ in the CNC rather than intermolecular BET from $CNC^{\bullet-}$ to $H_2P^{\bullet+}$ attached to the different CNC. The CS lifetime was determined from slope of the first-order plots to be 0.64(1) ms, which is the longest lifetime ever reported for electron donor-attached nanocarbon materials.[20] Such a long CS lifetime may result from the efficient electron migration in the CNT following the photoinduced charge separation.[20]

3.2.2 Multistep Charge Separation

A zinc porphyrin–C_{60} linked dyad (ZnP–C_{60}) is expanded by adding a ferrocene moiety (Fc) to ZnP to synthesize a Fc–MP–C_{60} triad ($M = Zn$ and H_2) and also to a Fc–ZnP–H_2P–C_{60} tetrad (Figure 3.9).[22] The BET (CR) rate constant in the dyad ($ZnP^{\bullet+}$–$C_{60}^{\bullet-}$) in the Marcus inverted region is much smaller than the CS rate constants of ET from both the singlet and triplet excited states of ZnP in the Marcus normal region (Figure 3.9).[22] The long CS lifetime enabled a subsequent ET from Fc to $ZnP^{\bullet+}$ in the triad (Fc–$ZnP^{\bullet+}$–$C_{60}^{\bullet-}$) and from ZnP to $H_2P^{\bullet+}$ in ZnP–$H_2P^{\bullet+}$–$C_{60}^{\bullet-}$ to produce the final CS state, Fc^+–ZnP–$C_{60}^{\bullet-}$ and $ZnP^{\bullet+}$–H_2P–$C_{60}^{\bullet-}$, in competition with the BET from $C_{60}^{\bullet-}$ to $ZnP^{\bullet+}$ in the initial CS states.[22] In the case of the CS state of the tetrad (Fc^+–ZnP–H_2P–$C_{60}^{\bullet-}$), in which charges are separated at a long distance ($R_{ee} = 48.9$ Å), the CS lifetime may be as long as that in the PRC.[22] The CR process in Fc^+–ZnP–H_2P–$C_{60}^{\bullet-}$ is located on the Marcus top region (Figure 3.9), when the CR rate constant is temperature independent (*vide infra*).[22]

Multistep charge-separation processes in Fc–ZnP–H_2P–C_{60} are shown in Scheme 3.2, where photoexcitation of Fc–ZnP–H_2P–C_{60} results in generation of the singlet excited state of ZnP (Fc–$^1ZnP^*$–H_2P–C_{60}), which has the highest energy (2.04 eV), followed by energy transfer from $^1ZnP^*$ to H_2P (k_{EN}) to produce Fc–ZnP–$^1H_2P^*$–C_{60}.[22] Then, ET from $^1H_2P^*$ to C_{60} occurs to produce the initial CS state (Fc–ZnP–$H_2P^{\bullet+}$–$C_{60}^{\bullet-}$), followed by subsequent ET from ZnP to $H_2P^{\bullet+}$ to produce the second CS state (Fc–$ZnP^{\bullet+}$–H_2P–$C_{60}^{\bullet-}$).[22] Finally, ET from the terminal electron donor (Fc) to $ZnP^{\bullet+}$ occurs to obtain the third (final) CS state (Fc^+–ZnP–H_2P–$C_{60}^{\bullet-}$) in which the plus and minus charges are separated a long distance ($R_{ee} = 48.9$ Å).[22]

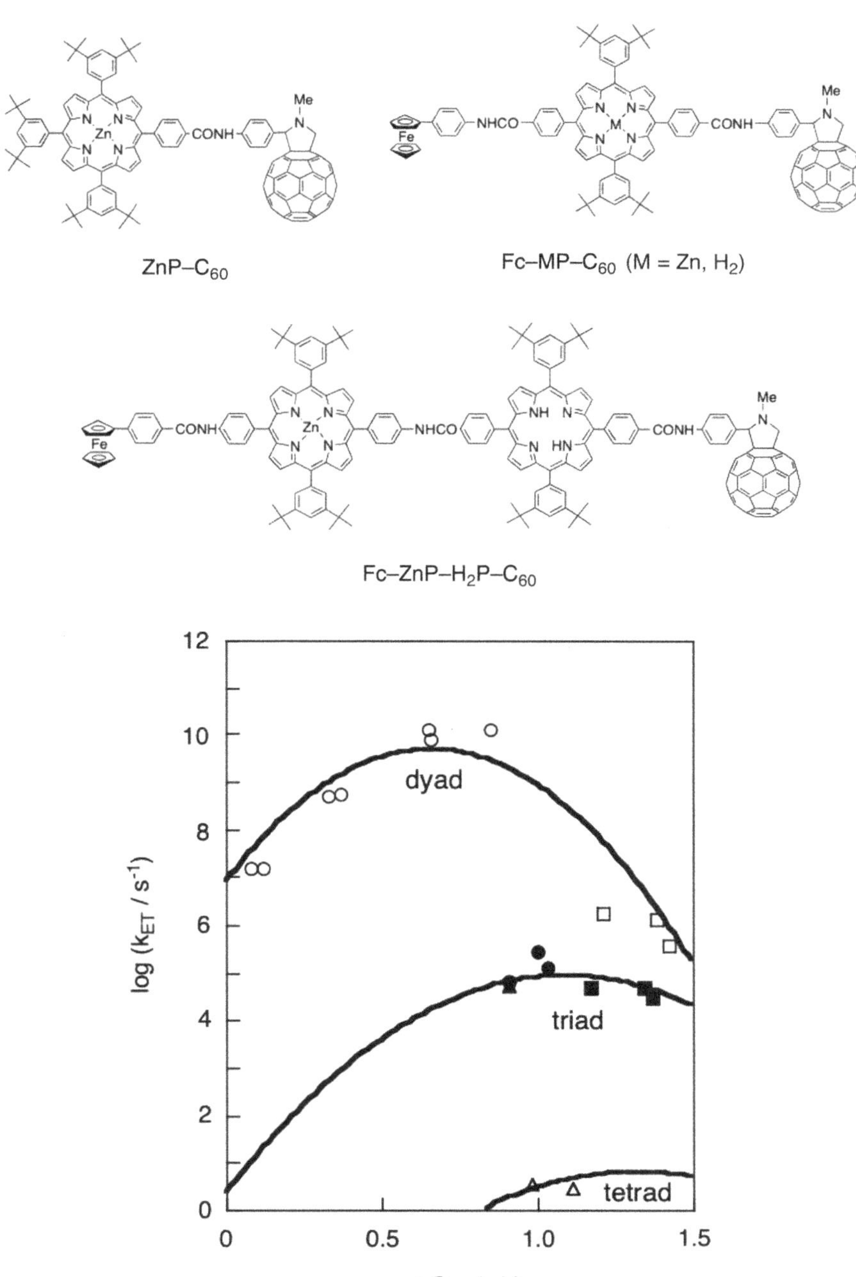

Figure 3.9 Driving force ($-\Delta G_{ET}$) dependence of the logarithm of intramolecular ET rate constants ($\log k_{ET}$) and BET ($\log k_{BET}$) in ZnP-C_{60} (CS: white circles; CR: white squares), Fc-ZnP-C_{60} (black circles), Fc-H_2P-C_{60} (black triangles), ZnP-H_2P-C_{60} (black squares), and Fc-ZnP-H_2P-C_{60} (white triangles). Reproduced from ref. 22 with permission from America Chemical Society, Copyright 2001.

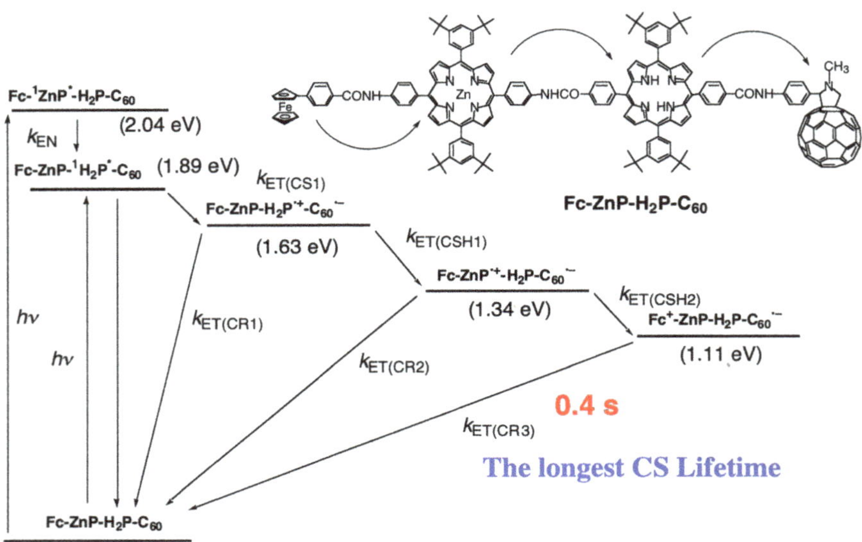

Scheme 3.2 Multistep charge separation processes in Fc–ZnP-H_2P-C_{60}.

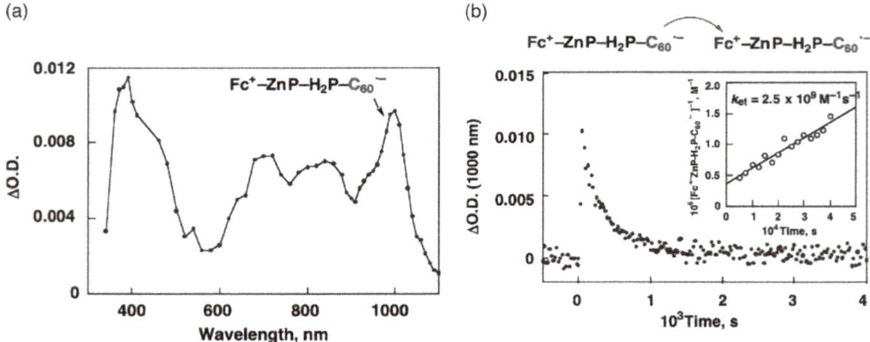

Figure 3.10 (a) Differential transient absorption spectrum observed upon nanosecond laser excitation at 532 nm of a nitrogen-saturated PhCN solution of Fc–ZnP-H_2P-C_{60} (7.2×10^{-6} M) with a time delay of 50 ns at 298 K. (b) The time profile of absorbance at 1000 nm due to the $C_{60}^{\bullet-}$ moiety in Fc^+-ZnP-H_2P-$C_{60}^{\bullet-}$ in argon-saturated DMF excited at 532 nm at 298 K. Inset: Second-order plot derived from the absorption change at 1000 nm. Adapted from ref. 22 with permission from America Chemical Society, Copyright 2001.

The CS state (Fc^+–ZnP–H_2P–$C_{60}^{\bullet-}$) was observed as a transient absorption spectrum upon laser excitation of a deaerated PhCN solution of Fc–ZnP–H_2P–C_{60} at 298 K (Figure 3.10a).[22] The CS state decays, obeying the second-order kinetics (Figure 3.10b), indicating that BET from the $C_{60}^{\bullet-}$ moiety to the Fc^+ moiety occurs *via* intermolecular ET (see the upper part of Figure 3.10b) rather than intramolecular ET, because the distance between the $C_{60}^{\bullet-}$ moiety in one Fc^+–ZnP–H_2P–$C_{60}^{\bullet-}$ molecule and the Fc^+ moiety in another Fc^+–ZnP–H_2P–$C_{60}^{\bullet-}$ molecule can be shorter compared with the long distance between the Fc^+ and $C_{60}^{\bullet-}$ moieties in the same molecule ($R_{ee} = 48.9$ Å).[22] The second-order rate constant of the bimolecular decay of the CS state was determined from the slope of the second-order plot (inset of Figure 3.10b) to be 2.5×10^9 M^{-1} s^{-1} in Ar-saturated DMF at 298 K, which is the same as the value of the diffusion rate constant in DMF.[22]

Formation of Fc^+–ZnP–H_2P–$C_{60}^{\bullet-}$ was also detected by EPR under photoirradiation of a deaerated PhCN solution of Fc–ZnP–H_2P–C_{60} at 203 K.[22] The EPR signal grows-in immediately upon 'turning on' the photoirradiation of Fc–ZnP–H_2P–C_{60}.[22] However, upon 'turning off' the photoirradiation, the EPR signal decayed slowly, obeying first-order kinetics to afford a rate constant of 2.8 s^{-1}, which corresponds to the CS lifetime of 0.38 s.[22] Under the frozen conditions, the diffusion was stopped to prohibit the intermolecular CR process, when only intramolecular BET from the $C_{60}^{\bullet-}$ moiety to the Fc^+ moiety Fc^+–ZnP–H_2P–$C_{60}^{\bullet-}$ was observed.[22] Under similar experimental conditions, the EPR response of Fc–ZnP–H_2P–C_{60} to on–off photoirradiation was much larger than that of ZnP–C_{60} a despite the lower quantum yield of the CS state of Fc–ZnP–H_2P–C_{60} ($\Phi = 0.24$) in PhCN relative to that of ZnP–C_{60} ($\Phi = 0.85$).[22]

Photoirradiation of similar tetrad and pentad compounds composed of Fc, ZnP and C_{60} afforded the CS states with the longest lifetime of 1.7 s.[23,24] The lifetime of the CS state remained the same irrespective of the change in temperature, because the BET (CR) process is located at the Marcus top region (Figure 3.9).[22] This is the first example among many PRC model compounds to achieve the CS lifetime, which is comparable to that observed for the PRC.[25–34] However, such a long CS lifetime could only be observed in frozen media, because the bimolecular BET between two Fc^+–ZnP–H_2P–$C_{60}^{\bullet-}$ molecules is much faster than the intramolecular BET in solution.[22] The maximum k_{ET} values (k_{ETmax}) of D–A dyads, triads and tetrads in the Marcus plot decreases with increasing the edge-to-edge distance (R_{ee}) of the CS state according to eqn (3.1), which is derived from eqn (2.20) and (2.21),[4]

$$\ln k_{ETmax} = \ln[2\pi^{3/2} V_0{}^2/(h(\lambda k_B)^{1/2})] - \beta R_{ee} \quad (3.1)$$

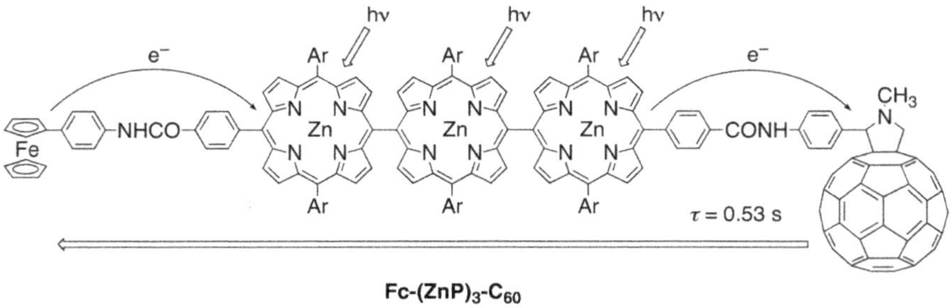

Fc-(ZnP)$_3$-C$_{60}$

Figure 3.11 Multistep photoinduced ET in a ferrocene-*meso,meso*-linked porphyrin trimer–fullerene pentad (Fc–(ZnP)$_3$–C$_{60}$); Ar = 3,5-^{t}Bu$_2$C$_6$H$_3$.

where V_0 refers to the electronic coupling element and β is the decay coefficient factor (damping factor), which depends primarily on the type of the bridging molecule (σ- or π-bonds). From the slope of the linear plot of ln k_{ETmax} *vs.* R_{ee}, the β value was determined to be 0.60 $\mathring{A}^{-1}$.[22] This β value with use of both σ- and π-bonds is between those of nonadiabatic ET for saturated hydrocarbon bridges with σ-bonds (0.8–1.0 $\mathring{A}^{-1}$) and unsaturated phenylene bridges with π-bonds (0.4 $\mathring{A}^{-1}$).[35]

A ferrocene-*meso,meso*-linked porphyrin trimer-fullerene linked pentad (Fc-(ZnP)$_3$–C$_{60}$ in Figure 3.11, where C$_{60}$ and Fc are tethered at both ends of a porphyrin trimer (ZnP)$_3$ (R_{ee} = 46.9 Å)) mimicked multistep ET processes in the photosynthetic reaction centre the best to afford the CS lifetime of 0.53 s at 163 K without lowering the CS efficiency (Φ = 0.83).[23]

3.2.3 Acridinium Ions

By connecting an electron donor, porphyrins and C$_{60}$ with covalent bonds, multistep photoinduced ET processes are made possible to mimic the function of the PRC to attain the long-lived CS state (*vide infra*). However, the CS state energy is limited because of relatively low-lying triplet excited states of porphyrins and fullerenes. In order to obtain high CS state energies, chromophores that have high lying triplet excited states and low reorganization energies are required as components of effective PRC models.

The ET reduction of 9-phenyl-10-methylacridinium ion (Acr$^+$–Ph) affords 9-phneyl-10-methylacridinyl radical (Acr$^•$–Ph) which shows many hyperfine splittings due to hyperfine coupling due to proton and nitrogen.[36] The hyperfine splitting (hfs) values of Acr$^•$–Ph were determined by computer simulation of the observed EPR spectrum.[36]

The linewidth of EPR signal of Acr•–Ph increases with increasing concentration of Acr^+–Ph because of the rapid electron-self exchange between Acr•–Ph and Acr^+–Ph.[36] The rate constant of electron-self exchange between Acr•–Ph and Acr^+–Ph was determined using eqn (2.24) from the slope of a plot of ΔH_{msl} *vs.* [Acr^+–Ph] to be $3.1 \times 10^9\ M^{-1}\ s^{-1}$ in MeCN at 298 K, which corresponds to $\lambda = 0.34$ eV with the use of eqn (2.16) where $\Delta G_{ET} = 0$.[36] The λ value decreases with decreasing the dielectric constant (polarity) and λ was as small as 0.21 eV in benzene at 298 K.[36] The activation entropy ($\Delta S^{\neq}$) was determined from the slope of the Eyring plot to be 0.5 cal K^{-1} mol^{-1}, which is close to zero.[36] Thus, the Acr^+ moiety may act an excellent candidate as a chromophore as well as an electron acceptor in electron donor–acceptor linked dyads. The high triplet excited energy of the acridinium ion is also suitable for a D–A dyad to afford the long-lived ET state with the CS energy that is higher than the energy of the triplet excited sate.[36]

Then, an electron donor moiety (mesityl group) is directly linked at the 9-position of the acridinium ion to synthesize 9-mesityl-10-methylacridinium ion (Acr^+–Mes), in which the solvent reorganization of ET is minimized because of the short linkage between the donor and acceptor moieties.[37] The X-ray crystal structure of Acr^+–Mes is shown in Figure 3.12a, where the dihedral angle between the two aromatic ring planes is perpendicular when there is no π conjugation or interaction between the donor and acceptor moieties.[37] In such a case, the UV-vis electronic absorption and emission (fluorescence) spectra of Acr^+–Mes are superpositions of

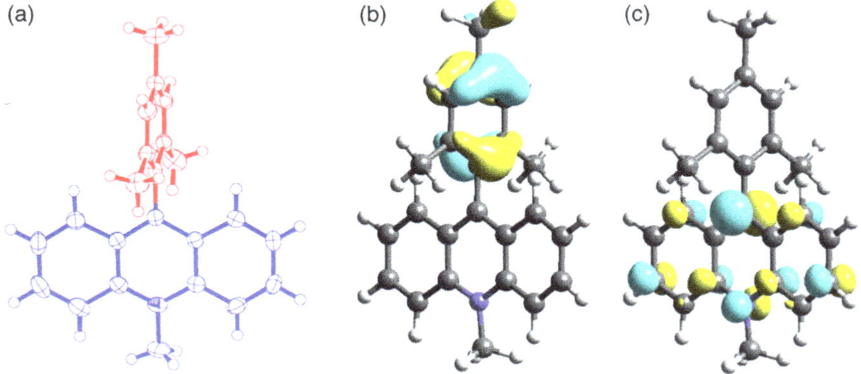

Figure 3.12 (a) X-ray crystal structure of Acr^+-Mes. (b) HOMO and (c) LUMO orbitals calculated by a DFT method with Gaussian 98 (B3LYP/6-31G* basis set). Reproduced from ref. 37 with permission from American Chemical Society, Copyright 2004.

that of each component. In addition, the HOMO and LUMO energies are completely localized on the Mes and Acr^+ moieties (Figure 3.12b and c), respectively.[37] The energy of the ET state ($Acr^•$–$Mes^{•+}$) in PhCN was determined from the E_{ox} (*vs.* SCE) value (1.88 V) of the Mes moiety and the E_{red} (*vs.* SCE) value of the Acr^+ moiety (−0.49 V) using the cyclic voltammograms to be 2.37 eV.[37] This is much higher than the CS energy of the PRC (0.5 eV). The rate constant of ET from the Mes moiety to the singlet excited state of Acr^+ moiety in $^1Acr^{+*}$–Mes was determined to be 2.4×10^{11} s^{-1} in PhCN at 298 K.[37] The decay rate of $Acr^•$–$Mes^{•+}$ in deaerated PhCN at 298 K obeyed second-order kinetics as in the case of Fc^+–ZnP–H_2P–$C_{60}{}^{•-}$ (Figure 3.10b) because the bimolecular BET between two $Acr^•$–$Mes^{•+}$ molecules is much faster than the intramolecular BET (*vide infra*).[37] At higher temperatures (*e.g.*, 383 K), however, the BET obeyed first-order kinetics (Figure 3.13a), indicating the intramolecular BET is highly temperature dependent due to the much larger activation energy of intramolecular BET than that of diffusion.[37] In frozen PhCN at 203–243 K, the first-order rate constant of intramolecular BET (Figure 3.13b) is also highly temperature dependent.[37] The overall temperature dependence of k_{BET} is fitted well by the Eyring plot, which affords $\lambda = 0.79$ eV (Figure 3.13c).[37]

Photoirradiation of a deaerated PhCN solution of Acr^+–Mes at 298 K resulted in no change in the absorption spectrum because of the bimolecular decay of $Acr^•$–$Mes^{•+}$.[38] When the photoirradiation was performed at low temperatures (213–243 K) with the use of a 1000 W high-pressure mercury lamp through the UV light cutting filter (>390 nm) and the irradiated sample was cooled to 77 K, the colour of the frozen sample at 77 K changed from green to brownish due to the formation of the ET state.[38] When a glassy 2-methyltetrahydrofuran (2-MeTHF) was employed for the photoirradiation of Acr^+–Mes at low temperature, the resulting glassy solution measured at 77 K showed the absorption spectrum due to ^{3}($Acr^•$–$Mes^{•+}$), consisting of the absorption bands due to the $Acr^•$ moiety (500 nm) and the $Mes^{•+}$ moiety (470 nm).[38] No decay of the absorption due to ^{3}($Acr^•$–$Mes^{•+}$) was observed until liquid nitrogen ran out as expected from the Eyring plot in Figure 3.13c.[38]

The long lifetime of the ET state of Acr^+–Mes has made it possible to observe the structural change in the Acr^+–Mes($ClO_4{}^-$) crystal upon photoirradiation with the use of a laser pump and X-ray probe crystallographic analysis.[39] Photoexcitation of the crystal of Acr^+–Mes($ClO_4{}^-$) resulted in a change in the bending angle of the *N*-methyl group of the Acr^+ moiety by 10.3(16)° when the *N*-methyl carbon moved 0.27(4) Å away from the mean plane of the ring due to the formation of the ET state, $Acr^•$–$Mes^{•+}$($ClO_4{}^-$).[39] This change

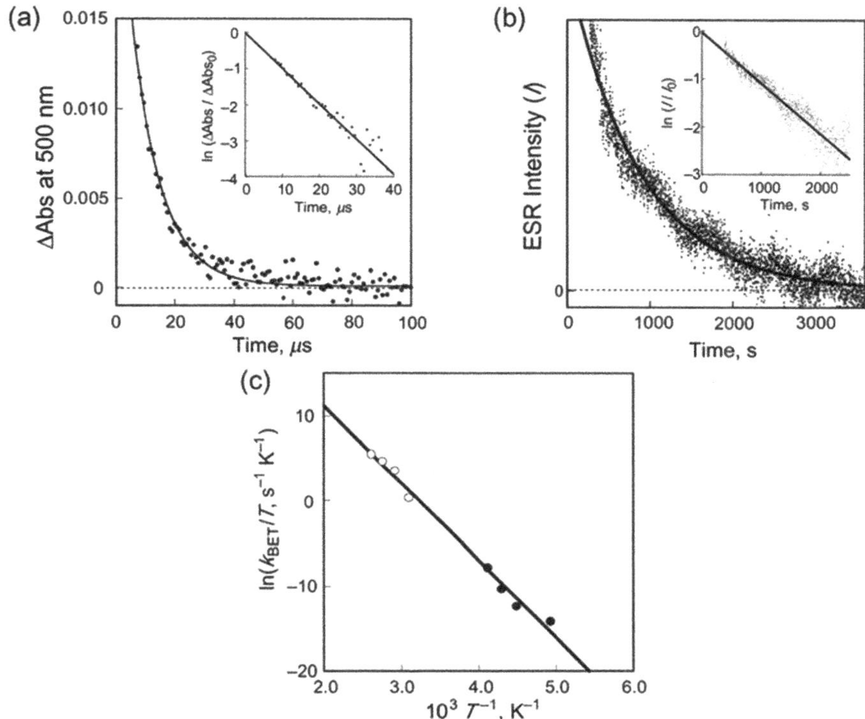

Figure 3.13 (a) Decay time profile of transient absorbance due to Acr•–Mes•+ in PhCN at 383 K. (b) Decay time profile of the EPR signal intensity due to Acr•–Mes•+ in frozen PhCN at 223 K. Insets: First-order plot of ln(I/I_0) *vs.* time. (c) Plot of ln (k_{BET}/T) *vs.* T^{-1} for intramolecular BET in Acr•–Mes•+ in PhCN determined by laser-induced transient absorption (○) and EPR (●) measurements. Reproduced from ref. 37 with permission from American Chemical Society, Copyright 2004.

results from ET from the Mes moiety to the singlet excited state of the Acr^+ moiety to produce the ET state (Acr•–Mes•+), in which the sp^2 carbon of the *N*-methyl group of the Acr^+ moiety before photoirradiation is changed to an sp^3 carbon in the Acr• moiety in the ET state.[39] The bending of the *N*-methyl group in Acr•–Mes•+ was accompanied by the rotation and movement of the ClO_4^- group by the electrostatic interaction with the Mes•+ moiety in Acr•–Mes•+.[39] Thus, the observed bending of the *N*-methyl group and the movement of ClO_4^- upon photoexcitation of the Acr^+–Mes(ClO_4^-) crystal provide strong evidence for the generation of the long-lived ET state, Acr•–Mes•+.[39] In contrast to Acr^+–Mes, no geometrical difference was observed by photoexcitation of Acr^+–Ph, which resulted in no ET from the Ph moiety to the $^1Acr^{+*}$ moiety.[39]

Immobilization of Acr^+–Mes has been achieved by cation exchange with a Acr^+–Mes cation in nanosized mesoporous silica–alumina (AlMCM-41) to prepare a Acr^+–Mes incorporated nanocomposite (Acr^+–Mes@AlMCM-41).[40] A tubular or rod-like (tAlMCM-41) morphology of diameter of 50–100 nm with length of 0.2–2 μm array and also a sphere morphology (sAlMCM-41) were obtained as revealed by TEM images (Figure 3.14).[40] The X-ray power pattern of tAlMCM-41 showed uniform channels with *ca.* 4 nm in diameter, which is large enough to accommodate Acr^+–Mes molecules by cation exchange upon mixing Na^+-exchanged AlMCM-41 with Acr^+–Mes in MeCN.[40] The cation exchange percentages of tAlMCM-41 and sAlMCM-41 by Acr^+–Mes were determined to be 16% and 18%, respectively.[40] The incorporated Acr^+–Mes into AlMCM-41 is stable without leaching out in MeCN at ambient temperature.[40]

Photoirradiation of Acr^+–Mes@tAlMCM-41 suspended in deaerated MeCN with the use of a 1000 W high-pressure mercury lamp through a UV-light cutting filter ($\lambda > 390$ nm) at 298 K results in formation of the ET state ($Acr^\bullet$–$Mes^{\bullet+}$) *via* photoinduced ET from the Mes moiety to the singlet excited state of the Acr^+ moiety.[40] The resulting EPR spectrum of $Acr^\bullet$–$Mes^{\bullet+}$ is composed of the superposition of the EPR spectra of $Acr^\bullet$–Mes and a mesitylene radical cation, although line broadening was observed due to spin–spin interaction.[40] The decay of the EPR signal intensity observed upon cutting off the light obeyed first-order kinetics, indicating the intramolecular BET from the $Acr^\bullet$ moiety to the $Mes^{\bullet+}$ moiety of $Acr^\bullet$–$Mes^{\bullet+}$ in MeCN at 298 K occurs with lifetime of 2.2 s instead of bimolecular second-order decay of two

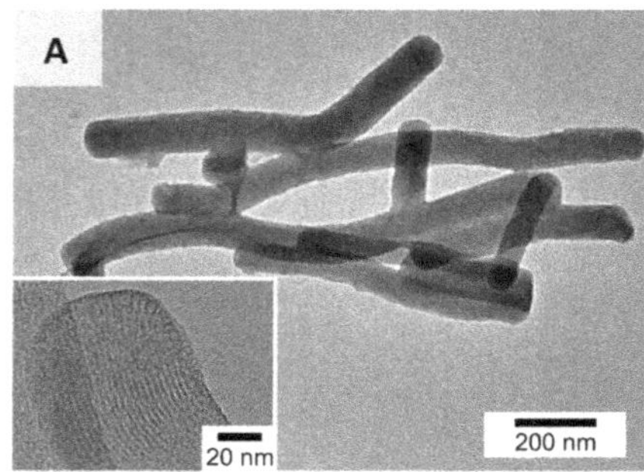

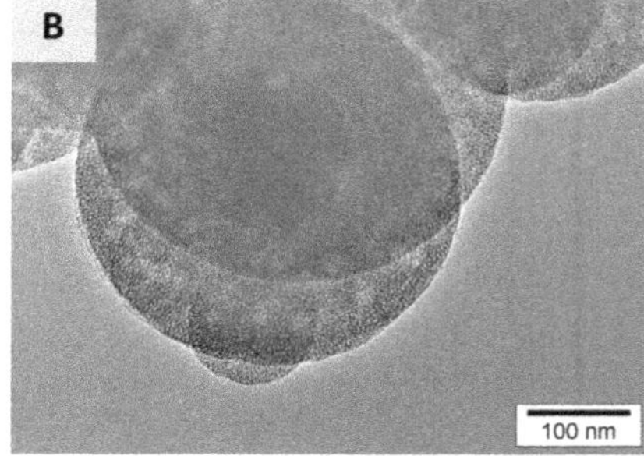

Figure 3.14 TEM images of (A) tubular-shaped nanosized mesoporous silica-alumina (tAlMCM-41) and (B) sphere-shaped nanosized mesoporous silica–alumina (sAlMCM-41). A high-resolution image of tAlMCM-41 is inserted in (A). Reproduced from ref. 40 with permission from the National Academy of Sciences of the United States of America, Copyright 2012.

$Acr^{\bullet}$–$Mes^{\bullet +}$ molecules, which was prohibited by immobilization of Acr^{+}–Mes inside AlMCM-41.[40] Thus, incorporation of Acr^{+}–Mes into tAlMCM-41 enabled elongating the lifetime of the ET state *via* one-step photoinduced ET, which is even longer than that of natural photosynthetic reaction centre *via* multistep charge separation (1 s).[8,40]

In order to determine the spin multiplicity of the ET state of $Acr^{\bullet}$–$Mes^{\bullet +}$@tAlMCM-41, an EPR spectrum was measured at 4 K in the dark after initial photoirradiation for a few seconds, exhibiting signals at $g = 2.0027$ signal with fine structure together with a strong sharp signal at $g = 4.0$, which clearly indicates the triplet multiplicity.[40] The zero-field splitting parameters D and E values were determined to be 59.9 G and 11.4 G, respectively.[40] By applying eqn (3.2),

$$D = 3\mu_0 g^2 \beta_e^{\,2}/(8\pi R^3) \tag{3.2}$$

where μ_0 is the permeability of a vacuum, g is 2.0027, R is the distance between two electron spins, and β_e is the Bohr magneton, R was estimated to be 7.7 Å, which agrees with the expected distance of 7.2 Å between an sp^2 carbon atom at the 4 position of the mesityl moiety and sp^2 carbon atoms at the 3 and 6 positions of the acridinyl moiety.[40]

A series of 9-substituted 10-methylacridinium ions (Acr^{+}–R in Figure 3.15) were synthesized to study photoinduced ET and BET.[41] The fluorescence lifetime of the unlinked 10-methylacridinium ion ($AcrH^{+}$: $\tau = 37$ ns in MeCN) was significantly reduced in Acr^{+}–1NA ($\tau = 310$ fs) due to quenching of the singlet excited state of the Acr^{+} moiety by the 1NA moiety.[41] Photoexcitation of Acr^{+}–1NA in MeCN with a femtosecond laser at 420 nm afforded a transient absorption spectrum with maxima at 530 and 700 nm, which are assigned to the acridinyl radical[42] and naphthalene radical cation,[43] respectively, in the singlet ET state $[^1(Acr^{\bullet}–1NA^{\bullet +})]$. The rate constant of formation of the singlet ET state (3.4×10^{12} s^{-1}) agrees with the fluorescence decay rate constant of Acr^{+}–1NA (3.2×10^{12} s^{-1}).[41] Such agreement of the rate constant of formation of the singlet ET state with the fluorescence decay rate constant was also confirmed for Acr^{+}–Mes in MeCN at 298 K.[41] The formation of the singlet ET state $[^1(Acr^{\bullet}–1NA^{\bullet +})]$ was followed by the intersystem crossing (ISC) to generate the triplet ET state with the rate constant of 2.5×10^8 s^{-1} in MeCN at 298 K.[41] In the case of Acr^{+}–1NA(2-Me), Acr^{+}–2NA and Acr^{+}–Dur as well, the triplet ET state $[^3(Acr^{\bullet}–R^{\bullet +})]$ was produced, but the quantum yield was quite changed depending on R.[41] In the case of

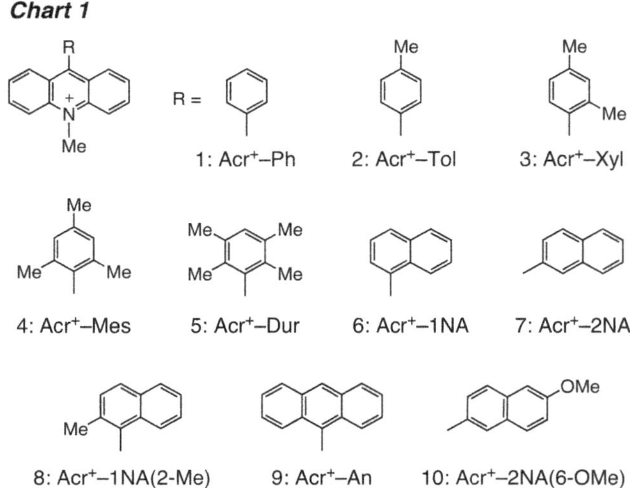

Figure 3.15 9-Substituted 10-methylacridinium ion derivatives (Acr^+-R). Reproduced from ref. 41 with permission from John Wiley and Sons, © 2017 Wiley-VCH Verlag GmbH & Co. KGaA, Weinheim.

Acr^+–Mes, the quantum yield is the largest (98%), whereas that of $^3(Acr^\bullet–Dur^{\bullet+})$ was only 4%.[41] In the $Dur^{\bullet+}$ moiety, there is pronounced spin distribution on 2- and 6-methyl groups, which can have orbital interaction with the $Acr^\bullet$ moiety in $Acr^\bullet–Dur^{\bullet+}$.[41] Such orbital interaction between the $Dur^{\bullet+}$ and $Acr^\bullet$ moieties of $Acr^\bullet–Dur^{\bullet+}$ would result in fast BET in the singlet ET state in competition with ISC to afford the long-lived triplet ET state.[41] The large Φ value (98%) of formation of $^3(Acr^\bullet–Mes^{\bullet+})$ may result from no orbital interaction between the $Acr^\bullet$ and $Mes^{\bullet+}$ moieties because of the orthogonal geometry between the two moieties and no spin density at the two *ortho* methyl groups of the $Mes^{\bullet+}$ moiety.[41] It should be noted that $^3(Acr^\bullet–Mes^{\bullet+})$ among $^3(Acr^\bullet–R^{\bullet+})$ is very special to obtain the high quantum yield as well as an extremely long lifetime.[41]

The decay of transient absorption due to $^3(Acr^\bullet–R^{\bullet+})$ obeyed second-order kinetics and the second-order decay rate constants were determined to always be close to the diffusion rate constants.[41] Thus, intramolecular BET from the $Acr^\bullet$ moiety to the $R^{\bullet+}$ moiety in $^3(Acr^\bullet–R^{\bullet+})$ is too slow to compete with intermolecular BET between two $^3(Acr^\bullet–R^{\bullet+})$ molecules.[41] The k_{BET} value of intramolecular BET from the $Acr^\bullet$ moiety to the $Mes^{\bullet+}$ moiety was determined by using Acr^+–Mes incorporated into nanosized mesoporous silica–alumina in MeCN (*vide supra*).[41]

Driving force dependence of the logarithm of the rate constant of ET from the R moiety to the singlet excited state of the Acr^+ moiety in Acr^+–R ($\log k_{ET}$) and BET from the $R^{\bullet+}$ moiety to the $Acr^{\bullet}$ moiety in $^3(Acr^{\bullet}–R^{\bullet+})$ is shown in Figure 3.16, the broken line was drawn based on eqn (2.20) with use of $\lambda = 0.92$ eV and $V = 330$ cm^{-1}.[41] The V value of 330 cm^{-1} is too large to treat vibrational motion classically. In addition, the V value for BET in $^3(Acr^{\bullet}–R^{\bullet+})$ to the singlet Acr^+–R should be much smaller than that for ET in the singlet excited state.[41] Thus, alternative understanding of the driving force dependence of $\log k_{ET}$ and $\log k_{BET}$ is provided by using semi-quantum theory using eqn (3.3),

$$k_{ET}^{q} = \frac{2\pi^{3/2}}{h}\frac{V^2}{\sqrt{\lambda_0 K_B T}}\sum_{i=0}^{\infty} e^{-s}\frac{S^i}{i!}\exp\left[-\frac{(\Delta G_{ET} + \lambda_0 + iE_v)^2}{4\lambda_0 k_B T}\right] \tag{3.3}$$

where E_v is the vibrational frequency, S is the electronic-vibrational coupling, and λ_o is the outer sphere reorganization energy.[44] The internal reorganization energy can be given by $\lambda_l = E_v/S$.[44] The fitting of the data of $\log k_{ET}$ and k_{BET} of the singlet states is shown in Figure 3.16 by the green dashed line with use of $V = 130$ cm^{-1}, $\lambda_o = 0.48$ eV, $E_v = 0.17$ eV, and $S = 1.9$.[40] The corresponding internal reorganization energy is $\lambda_l = 0.32$ eV and the total reorganization

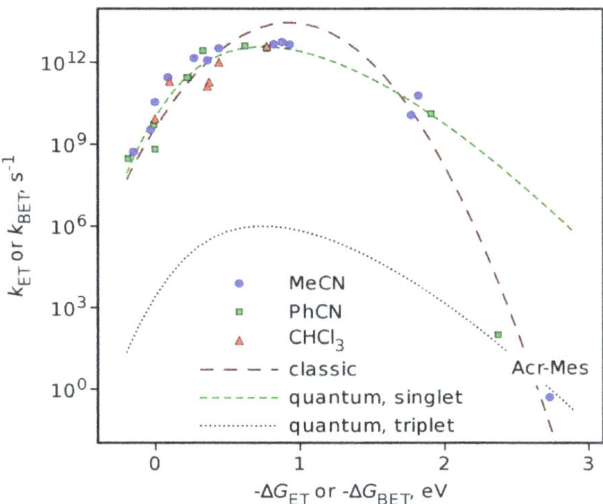

Figure 3.16 Driving force ($-\Delta G_{ET}$ or $-\Delta G_{BET}$) dependence of the logarithm of intramolecular ET and BET rate constants ($\log k_{ET}$ or $\log k_{BET}$) in Acr^+–R in MeCN (•), PhCN (▫) and $CHCl_3$ (▴) at 298 K. Reproduced from ref. 41 with permission from John Wiley and Sons, © 2017 Wiley-VCH Verlag GmbH & Co. KGaA, Weinheim.

energy $\lambda = \lambda_o + \lambda_i = 0.8$ eV.[41] The k_{BET} values of the triplet ET states $^3(Acr^\bullet–R^{\bullet+})$ (two points in the right-bottom corner of the plot) are much smaller than the fitting for the singlet states.[41] Assuming that only V differs in singlet and triplet ET states, the dotted line fitting the k_{BET} values of the triplet ET states predicts the electronic coupling to be $V_T = 0.07$ cm^{-1}, which is much smaller than that for the singlet ET (CS and CR).[41] However, the temperature dependence of k_{BET} (Figure 3.13c) fits well with eqn (2.20) rather than eqn (3.3).[41] Thus, the highly temperature dependent long lifetime of the triplet ET state $[^3(Acr^\bullet–R^{\bullet+})]$ results from the BET process deeply in the Marcus inverted region rather than the small V value of the triplet ET state.[41]

The strong oxidizing and reducing capability of $^3(Acr^\bullet–Mes^{\bullet+})$ has been confirmed by intermolecular ET reactions from various electron donors (*e.g.*, naphthalene and anthracene) to the $Mes^{\bullet+}$ moiety of $^3(Acr^\bullet–Mes^{\bullet+})$ and those from the $Acr^\bullet$ moiety to various electron acceptors (*e.g.*, *p*-benzoquinone and hexyl viologen).[41] The dependence of the logarithm of the rate constants of ET reactions of $^3(Acr^\bullet–Mes^{\bullet+})$ ($\log k_{et}$) on the ET driving force ($-\Delta G_{et}$) is shown in Figure 3.17, where the best fit line based on eqn (2.16) for the outer-sphere ET affords $\lambda = 1.0$ eV.[41]

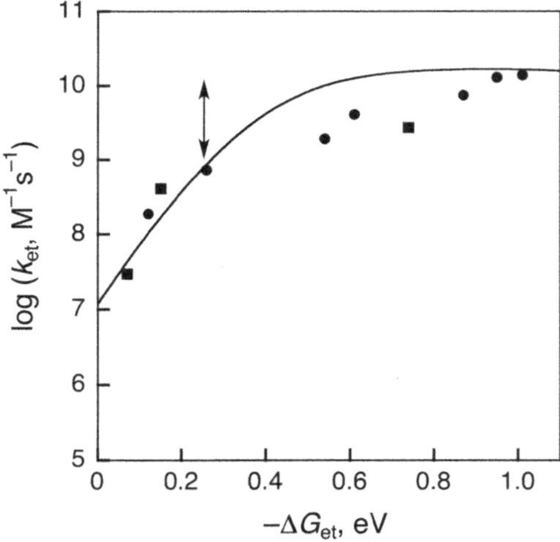

Figure 3.17 Driving force ($-\Delta G_{et}$) dependences of $\log k_{et}$ for intermolecular ET from electron donors (●) to the $Mes^{\bullet+}$ moiety and from the $Acr^\bullet$ moiety in the triplet ET state of $Acr^\bullet–Mes^{\bullet+}$ to electron acceptors (■) in MeCN at 298 K, and the fitting of the line based on the Marcus theory of outer-sphere ET [eqn (2.16)]. Reproduced from ref. 41 with permission from John Wiley and Sons, © 2017 Wiley-VCH Verlag GmbH & Co. KGaA, Weinheim.

In order to improve the oxidizing ability of the $Mes^{\bullet +}$ moiety of $^3(Acr^{\bullet}-Mes^{\bullet +})$, 9-(4-X-2,6–dimethylphenyl)-10-methylacridinium ions (Acr^+–XylX, X = F and Cl) were synthesized and the photodynamics was examined.[44] The X-ray crystal structures of Acr^+–XylX (X = F and Cl) showed the dihedral angle made by the two aromatic ring planes was approximately perpendicular (80° for Acr^+–XylF and 79° for Acr^+–XylCl) as reported for Acr^+–Mes (Figure 3.13a).[37] There is no π-conjugation between the donor and acceptor moieties.[44] Thus, the absorption and fluorescence spectra of Acr^+–XylX are superpositions of those of each component, *i.e.*, 4-chloro-2,6-xylenyl (XylCl) and 10-methylacridinium ion (Acr^+) moieties.[44] The one-electron reduction potentials (E_{red} *vs.* SCE) of Acr^+–XylCl and Acr^+–XylF in deaerated MeCN were determined to be the same, −0.53 V.[44] The one-electron oxidation potentials (E_{ox} *vs.* SCE) of Acr^+–XylF and Acr^+–XylCl were also determined to be +2.20 V and +2.21 V, respectively.[44] The E_{ox} value of Acr^+–XylCl is higher than the E_{ox} values of Acr^+–Mes (+2.06 V) and toluene (+2.20 V).[44] The higher E_{ox} value of Acr^+–XylCl than toluene enabled ET from toluene to the $XylCl^{\bullet +}$ moiety of $^3(Acr^{\bullet}-XylCl^{\bullet +})$ produced upon photoexcitation of Acr^+–XylCl, resulting in photocatalytic oxygenation of toluene by dioxygen.[44]

The photocatalytic trifluoromethylation of toluene derivatives has also been made possible by using Acr^+–XylCl as a photoredox catalyst and *S*-(trifluoromethyl)-dibenzothiophenium (CF_3DBT^+) as a trifluoromethyl source (Scheme 3.3).[44–46] The photocatalytic trifluoromethylation is started by intramolecular ET from the XylCl moiety to the singlet-excited state of the Acr^+ moiety of Acr^+–XylCl to produce $^3(Acr^{\bullet}-XylCl^{\bullet +})$, followed by ET from toluene to the $XylCl^{\bullet +}$ moiety of $^3(Acr^{\bullet}-XylCl^{\bullet +})$ to produce a toluene radical cation and $Acr^{\bullet}$–Xyl.[44] On the other hand, ET from $Acr^{\bullet}$–Xyl to CF_3DBT^+ occurs to produce a $CF_3^{\bullet}$ radical, which is added to toluene to produce the CF_3 radical adduct that is oxidized by a toluene radical cation to yield the trifluromethylated product, accompanied by deprotonation.[44]

The high quantum yield of generation of $^3(Acr^{\bullet}-Mes^{\bullet +})$, which has the high oxidizing and reducing ability with an extremely long lifetime (*vide infra*), has made Acr^+–Mes one of the best photoredox catalysts utilized in a variety of photocatalytic synthetic applications.[14,31,44,47–58]

3.2.4 Quinolinium Ions

The Acr^+ moiety of Acr^+–1NA was replaced by a $QuPh^+$ moiety to synthesize 2-phenyl-4-(1-naphthyl)quinolinium ion ($QuPh^+$–NA) in

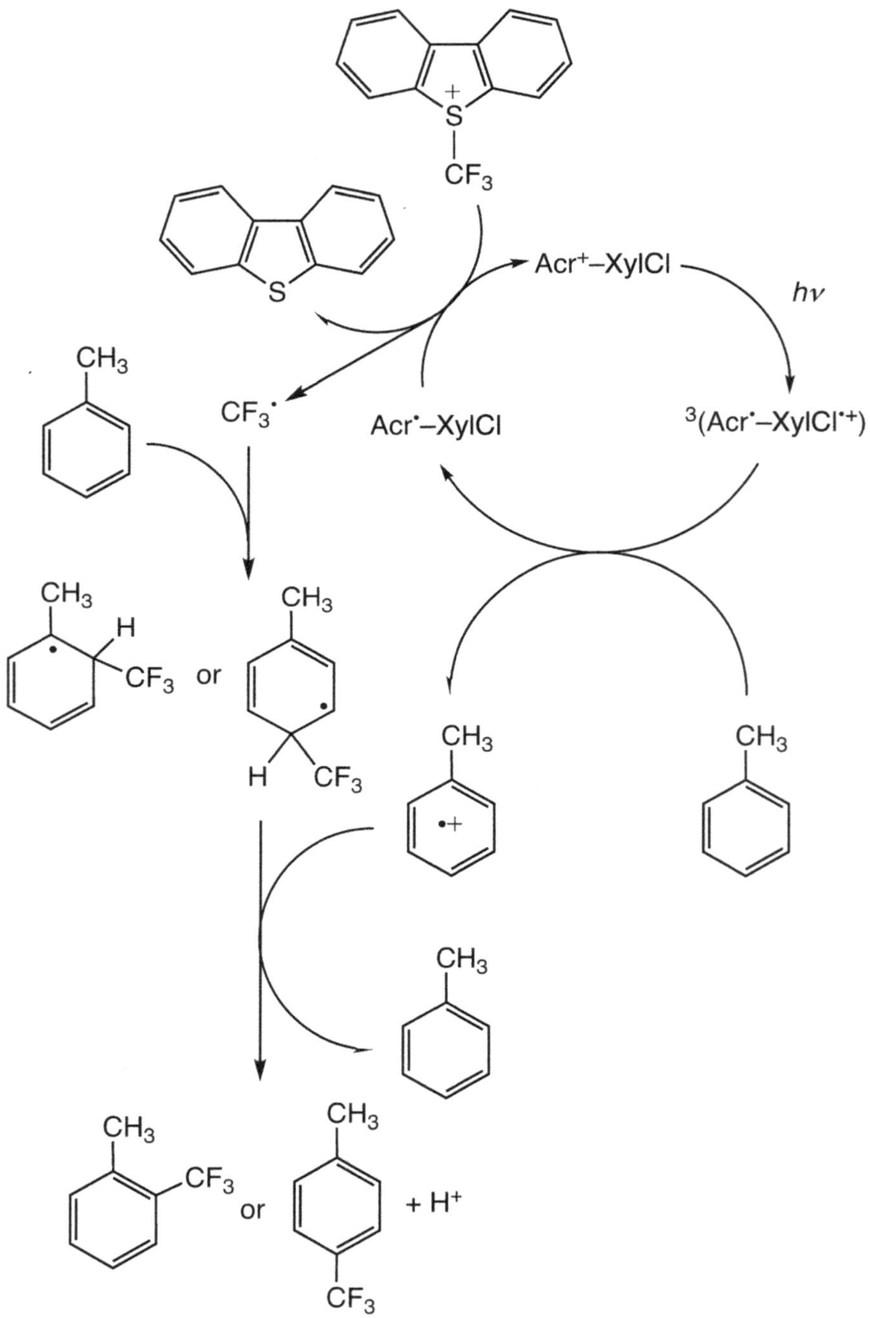

Scheme 3.3 Mechanism of photocatalytic trifluoromethylation with Acr^{+}–XylCl *via* the ET oxidation of toluene. Reproduced from ref. 44 with permission from American Chemical Society, Copyright 2024.

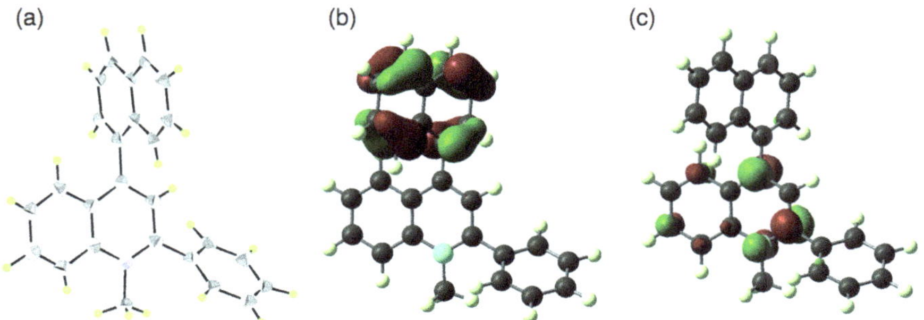

Figure 3.18 (a) ORTEP drawing of the X-ray crystal structure of $QuPh^+$-NA. (b) HOMO and (c) LUMO of $QuPh^+$-NA obtained by DFT calculations with use of B3LYP/6-31G basis set. Reproduced from ref. 59 with permission from the Royal Society of Chemistry.

which a naphthalene (NA) donor moiety is directly linked with a quinolinium ion (QuH^+).[59] The X-ray crystal structure of $QuPh^+$–NA in Figure 3.18a showed that the dihedral angle was nearly perpendicular (87°) between the NA and Qu^+ moieties of $QuPh^+$–NA due to steric interaction with the hydrogen at the peri-position.[59] The HOMO and LUMO orbitals of $QuPh^+$–NA are localized on the NA and Qu^+ moieties, respectively (Figure 3.18 parts b and c, respectively).[59] Femtosecond laser excitation at 390 nm of a deaerated MeCN solution of $QuPh^+$–NA resulted in formation of the singlet ET state [^{1}(QuPh$^\bullet$–NA$^{\bullet+}$)] within 0.5 ps, followed by rapid ISC to produce the triplet ET state with a quantum yield of 83%.[59] The triplet ET state decayed *via* intermolecular BET with the second-order rate constant of 3.3×10^9 M^{-1} s^{-1} in deaerated MeCN at 298 K.[59]

The E_{red} value (*vs.* SCE) of the $QuPh^+$ moiety of $QuPh^+$–NA (−0.90 V) is much more negative than that of the Acr^+ moiety of Acr^+–Mes (−0.57 V),[60] indicating that the ET state (QuPh$^\bullet$–NA$^{\bullet+}$) acts as a much stronger reductant than that of Acr$^\bullet$–Mes$^{\bullet+}$.[59] In fact, the rate constant (k_{et}) of ET from the QuPh$^\bullet$ moiety of the triplet ET state [^{3}(QuPh$^\bullet$–NA$^{\bullet+}$)] to hexyl viologen (HV^{2+}) was determined to be 1.3×10^9 M^{-1} s^{-1}, which is much larger than the corresponding k_{et} value of ET from the Acr$^\bullet$ moiety of ^{3}(Acr$^\bullet$–Mes$^{\bullet+}$) to HV^{2+} (1.0×10^8 M^{-1} s^{-1}).[59] When HV^{2+} was replaced by tetracyanobenzene (TCNB: E_{red} *vs.* SCE = −0.74 V),[42] ET from the QuPh$^\bullet$ moiety of ^{3}(QuPh$^\bullet$–NA$^{\bullet+}$) to TCNB occurred efficiently.[59] In contrast to the case of ^{3}(QuPh$^\bullet$–NA$^{\bullet+}$), no ET from the Acr$^\bullet$ moiety of ^{3}(Acr$^\bullet$–Mes$^{\bullet+}$) to TCNB occurred, because the ET from the Acr$^\bullet$ moiety to TCNB is endergonic ($\Delta G_{et} = +0.17$ eV).[59]

As in the case of Acr$^+$–Mes, QuPh$^+$–NA was also immobilized inside nanosized mesoporous silica–alumina with a spherical shape (sAlMCM-41) by cation exchange to produce the nanocomposite (QuPh$^+$–NA@sAlMCM).[61] The TEM image of QuPh$^+$–NA@sAlMCM revealed porous hollow sphere particles with diameters of 200–800 nm.[61] The diffused reflectance UV-vis spectrum of the QuPh$^+$–NA@sAlMCM-41 composite (red solid line) is similar to peaks and shoulders of QuPh$^+$–NA ion in MeCN.[61] It was shown that 7.9% of all the Na$^+$-sites were occupied by QuPh$^+$–NA.[61]

Photoirradiation of the QuPh$^+$–NA@sAlMCM-41 composite with use of a 1000 W high-pressure mercury lamp through a UV-light cutting filter ($\lambda > 340$ nm) resulted in formation of the triplet ET state [^{3}(QuPh$^\bullet$–NA$^{\bullet+}$)@sAlMCM-41] *via* ET from the naphthalene (NA) moiety to the singlet excited state of the quinolinium ion (QuPh$^+$) moiety.[61] The decay of the EPR signal intensities observed by cutting off the light obeyed first-order kinetics with a lifetime longer than 190 s at 316 K.[61] The activation enthalpy ($\Delta H^{\neq}$) of the rate constant of intramolecular BET in ^{3}(QuPh$^\bullet$–NA$^{\bullet+}$) was determined to be 12(2) kcal mol^{-1}.[61] A large $\Delta H^{\neq}$ value [20 (1) kcal mol^{-1}] has also been reported for Acr$^+$–Mes immobilized inside AlMCM-41 in MeCN,[40] resulting from a large driving force for the BET, which is deeply in the Marcus inverted region, because of decreased solvation in the pores of sAlMCM-41.[61] The QuPh$^+$–NA@sAlMCM-41 composite, which was highly dispersed in water, was reported to act as an efficient and robust photoredox catalyst for the reduction of O_2 by oxalate to produce hydrogen peroxide with a quantum yield of 10% (Scheme 3.4).[61]

The long lifetime of ^{3}(QuPh$^\bullet$–NA$^{\bullet+}$) and the high reducing capability of the QuPh$^\bullet$ moiety has enabled efficient H_2 evolution using metal nanoparticles (MNPs: M = Pt, Ru and Ni) as H_2 evolution catalysts and QuPh–NA$^+$ as a photoredox catalyst under highly basic conditions (pH $\approx$ 10).[62] The photocatalytic cycle for H_2 evolution with

Scheme 3.4 Photocatalytic cycle for H_2O_2 production by O_2 reduction by oxalate using QuPh$^+$–NA@sAlMCM-41 as a photoredox catalyst. Reproduced from ref. 61 with permission from the Royal Society of Chemistry.

$QuPh^+$–NA is started by photoexcitation of $QuPh^+$–NA, which resulted in formation of $^3(QuPh^•–NA^{•+})$, followed by ET from NADH to the $NA^{•+}$ moiety to produce $NADH^{•+}$ and $QuPh^•$–NA.[62] $NADH^{•+}$ undergoes deprotonation to produce $NAD^•$, which is rapidly oxidized by $QuPh^+$–NA to produce NAD^+ and $QuPh^•$–NA.[62,63] Thus, two equivalents of $QuPh^•$–NA are produced by photoinduced ET reduction of $QuPh^+$–NA by NADH.[62] The rapid electron injection by $QuPh^•$–NA to MNPs results in efficient H_2 evolution under highly basic conditions (pH ≈ 10).[62] The photocatalytic system with RuNPs and $QuPh^+$–NA exhibited the highest H_2 evolution rate of 28 μmol h^{-1} from 2 mL of solution (14 μmol h^{-1} mL^{-1}) with 12.5 mg L^{-1} RuNPs at pH 10.[62] NiNPs are also effective for photocatalytic H_2 evolution, but the catalytic reactivity was less than that of PtNPs and RuNPs, although the catalytic reactivity increased with decreasing the size of NiNPs.[64]

3.3 Supramolecular Electron Transfer

3.3.1 π–π Interactions

The natural photosynthesis is composed of photosystem I (PSI) and photosystem II (PSII). In each photosystem, components of the photosynthetic reaction centre (PRC) are located at the appropriate position by noncovalent bonding in the protein matrix to optimize the charge-separation efficiency. In addition, light harvesting units are combined with the PRC by noncovalent bonding. The use of noncovalent bonding, such as π–π interaction, electrostatic (Coulombic) interaction and hydrogen bonds, has attracted significant attention as a simpler but more elegant way to construct D–A ensembles mimicking the efficient biological ET systems.[65–73]

In particular, π–π interaction plays an important role in biological systems, such as π-stacking of double-strand DNA. A simple example to form an electron donor–acceptor complex by π–π interaction has been reported for a π complex of a freebase bisporphyrin (H_4DPOx) with acridinium perchlorate ($AcH^+ClO_4^-$), as shown in Scheme 3.5. Formation of the π complex between H_4DPOx and AcH^+ was examined by the UV-vis and NMR spectral changes.[74] The formation constant between H_4DPOx and AcH^+ was determined form the spectral titration to be 9.7×10^4 M^{-1} in PhCN at 298 K.[74] The optimized structure of the π complex between H_4DPOx and AcH^+ obtained by DFT calculation indicated that $AcrH^+$ was inserted between the two porphyrin planes by π–π interaction. The HOMO (highest occupied

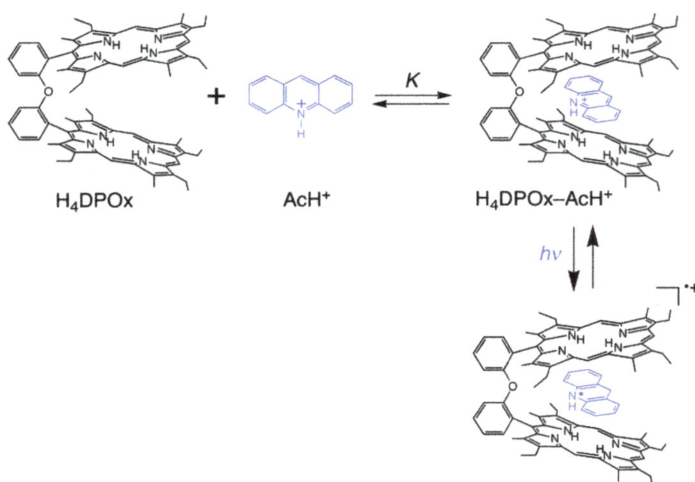

Scheme 3.5 Formation of a π complex between a freebase bisporphyrin (H_4DPOx) with acridinium perchlorate ($AcH^+ClO_4^-$) and photoinduced ET. Reproduced from ref. 74 with permission from American Chemical Society, Copyright 2006.

molecular orbital) and LUMO (lowest unoccupied molecular orbital) of the π complex are localized on the H_4DPOx and AcH^+ moieties, respectively.[74] Thus, photoexcitation of the π complex results in ET from the singlet excited state ($^1H_4DPOx^*$) to AcH^+ in the π complex to produce the ET state ($H_4DPOx^{\bullet +}$–$AcH^{\bullet}$) with the quantum yield of 90% (Scheme 3.5).[74] The rate constant of ET from $^1H_4DPOx^*$ to AcH^+ was determined by the time-resolved fluorescence measurements to be 2.9×10^8 s^{-1} in PhCN at 298 K.[74] The decay of transient absorption at 460, 520 and 650 nm due to the ET state ($H_4DPOx^{\bullet +}$–$AcH^{\bullet}$) obeyed first-order kinetics to afford the rate constant of BET to the ground state (5.5×10^4 s^{-1}) in PhCN at 298 K.[74] The corresponding lifetime of the ET state was determined to be 19 μs in PhCN at 298 K.[74] The lifetime of the ET state was highly temperature dependent and the temperature dependence of the rate constant of BET (k_{BET}) agreed well with eqn (3.4), which is derived from the Marcus equation of intramolecular nonadiabatic ET [eqn (2.20)], showing a linear correlation between $\ln(k_{BET}T^{1/2})$ *vs.* T^{-1}.[74] From the slope and intercept the λ and V values were determined to be 0.54 ± 0.1 eV and 1.6 ± 0.3 cm^{-1}, respectively.[74] The lifetime of the ET state at 77 K is expected to be as long as 360 days form the extrapolation of the linear plot of ln $(k_{BET}T^{1/2})$ *vs.* T^{-1}.[74] In fact, such an extremely long-lived ET state was observed by photoirradiation of a glassy 2-MeTHF solution containing 10% butyronitrile of the H_4DPOx–AcH^+ complex by use of a 1000 W high-pressure Hg lamp at low temperature.[74] The new absorption bands due to $H_4DPOx^{\bullet+}$ (450–540 nm and 620–800 nm) and $AcH^{\bullet}$ (450–540 nm) were clearly observed and the absorption bands due to the ET state exhibited no decay in 200 min at 77 K.[74] The colour changed from that of the ground state of the $H_4DPOx–AcH^+$ complex to that of the ET state produced by photoirradiation.[74] The colour and the absorption spectrum due to the ET state was returned back to those of the ground state π complex, when temperature was increased to 298 K and such a change could be repeated many times.[74]

$$\ln(k_{BET}T^{1/2}) = \ln(2\pi^{3/2}V^2/[h(\lambda k_B)^{1/2}]) - (\Delta G_{BET} + \lambda)^2/(4\lambda k_B T) \quad (3.4)$$

The driving force ($-\Delta G_{ET}$) dependence of the ET rate constant in the $H_4DPOx–AcH^+$ complex at 298 K was fitted well using eqn (2.20) with the λ and V values, which were determined from the temperature dependence of k_{BET}.[74] The k_{ET} value of ET from $^1H_4DPOx^*$ to AcH^+ in the $H_4DPOx–AcH^+$ complex (2.9×10^8 s^{-1}) agrees with the predicted value from the Marcus line within the experimental error.[74] The k_{ET} value is located nearly on the Marcus top region, whereas the k_{BET} value is located is deeply in the Marcus inverted region.[75] This is the reason why the lifetime of the ET state ($H_4DPOx^{\bullet+}–AcH^{\bullet}$) is extremely long due to the very slow BET at low temperature.[74]

The energy diagram of photoinduced ET in the $H_4DPOx^{\bullet+}–AcH^{\bullet}$ complex is shown in Scheme 3.6.[74] The energy level of the ET state ($-\Delta G_{ET}$) is determined to be 1.28 eV from the one-electron redox potentials of the $H_4DPOx–AcH^+$ complex, which is lower than the that

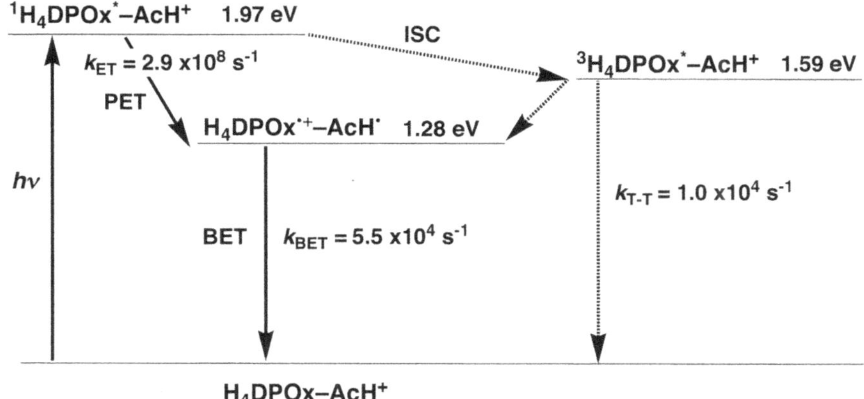

Scheme 3.6 Energy diagram of photoinduced ET and BET in the $H_4DPOx–AcH^+$ complex. Reproduced from ref. 74 with permission from American Chemical Society, Copyright 2006.

of the triplet states of H_4DPOx (1.59 eV) and AcH^+ (2.01 eV).[74] This is the reason why the transient absorption spectra due to the ET state can be observed and no T–T absorption due to the triplet excited states was observed.[74]

As in the case of the $H_4DPOx–AcH^+$ complex, a π-extended viologen, 1,4-bis(*N*-hexyl-4-pyridinium)butadiene dication (BHV^{2+}) is inserted between the two porphyrin moieties of the cofacial porphyrin dimer [H_4(DPA)], as shown in Scheme 3.7, where photoexcitation of the $H_4(DPA)–BHV^{2+}$ complex resulted in ET from $^1H_4(DPA)^*$ to BHV^{2+} to produce the ET state ($H_4(DPA)^{\bullet+}–BHV^{\bullet+}$) with quantum yield of 83% and lifetime of 590 μs in PhCN at 298 K.[76]

As shown above, electron acceptor guest molecules can be inserted between two porphyrin rings of porphyrin dimers with appropriate linkers by π–π interactions to form sandwich complexes.[75,77–82] A three dimensional π-conjugated molecule such as C_{60} can also be inserted between two porphyrins. Thus, C_{60} was inserted between the two porphyrin rings of a cyclic Ni porphyrin dimer (Ni–CPD_{Py}) linked by butadiyne moieties bearing 4-pyridyl groups to form a sandwich complex ($C_{60}\subset Ni_2–CPD_{Py}$).[83]

The X-ray crystal structure of $C_{60}\subset Ni_2–CPD_{Py}$ showed that one C_{60} molecule was inserted by tilting the porphyrin rings with respect to each other due to strong π–π interactions between the porphyrin rings and C_{60}.[83] The adjacent dimers are linked by hydrogen bonds and π–π interactions to arrange C_{60} molecules linearly in the inner channel of a supramolecular peapod.[83]

Photoexcitation of $C_{60}\subset Ni_2–CPD_{Py}$ resulted in no formation of the CS state, because the energy level of the CS state (1.98 eV) is higher than that of $^3C_{60}^*$ (1.60 eV).[83] Thus, the singlet excited state of the nickel porphyrin immediately changed to the triplet excited state by ISC to afford the low energy triplet excited state of C_{60} ($^3C_{60}^*$) without occurrence of ET.[83] When $Ni_2–CPD_{Py}$ was replaced by a free base porphyrin dimer ($H_4–CPD_{Py}$), however, the formation of the CS state

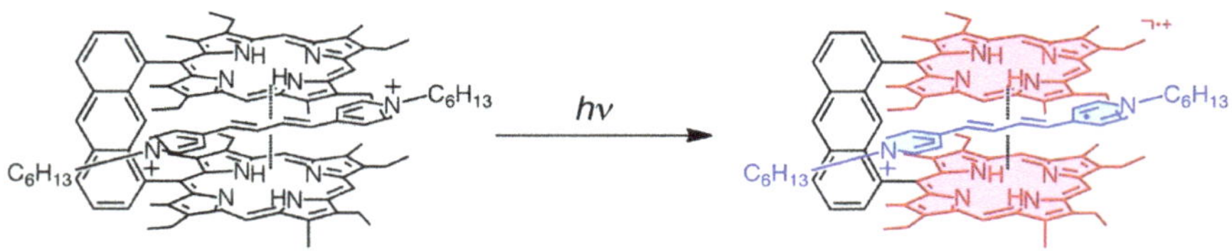

Scheme 3.7 Photoinduction in the supramolecular complex of a cofacial porphyrin dimer [H_4(DPA)] with a 1,4-bis(*N*-hexyl-4-pyridinium)butadiene dication (BHV^{2+}). Reproduced from ref. 76 with permission from the Royal Society of Chemistry.

[H_4-$CPD_{Py}^{\bullet+}$-$C_{60}^{\bullet-}$] was observed by femtosecond laser-induced transient absorption measurements of $C_{60}{\subset}H_4$-CPD_{Py} in the solid state with a lifetime of 470 ps.[84]

When C_{60} was replaced by Li^+ endohedral fullerene ($Li^+@C_{60}$), porphyrin dimers with four long alkoxy substituents on the *meso*-phenyl groups (MCPDPy(OC_6) in Figure 3.19) form strong supramolecular complexes in even a polar solvent such as PhCN, because $Li^+@C_{60}$ is a much stronger electron acceptor due to Li^+ encapsulation.[85] The formation constants (K) of $Li^+@C_{60}{\subset}MCPD_{Py}(OC_6)$ in PhCN at 298 K were determined by spectral titration to be 2.6×10^5 M^{-1} for $Li^+@C_{60}{\subset}H_4$-$CPD_{Py}(OC_6)$ and 3.5×10^5 M^{-1} for $Li^+@C_{60}{\subset}Ni_2$-$CPD_{Py}(OC_6)$.[85]

Laser photoexcitation of $Li^+@C_{60}{\subset}Ni_2$-$CPD_{Py}(OC_6)$ resulted in ET from Ni_2-$CPD_{Py}(OC_6)$ to the triplet excited state of $Li^+@C_{60}$ to produce the CS state [$Li^+@C_{60}^{\bullet-}{\subset}Ni_2$-$CPD_{Py}(OC_6)^{\bullet+}$], which exhibited transient absorption bands due to $Li^+@C_{60}^{\bullet-}$ at 1035 nm.[85] The decay of the transient absorption band at 750 nm due to $^3Li^+@C_{60}^*$ coincided with the rise in the absorption band at 1035 nm due to $Li^+@C_{60}^{\bullet-}$.[85] The rate constant of ET from Ni_2-$CPD_{Py}(OC_6)$ to $^3Li^+@C_{60}^*$ to produce the CS state was determined from the rise in absorbance at 1035 nm to be 5.7×10^7 s^{-1}.[85] The decay of absorbance at 1035 nm due to $Li^+@C_{60}^{\bullet-}$ obeyed first-order kinetics with the same slope irrespective of the difference in the laser power, indicating that the CS decayed *via* intrasupramolecular BET rather than a bimolecular BET between the two CS states.[85] The CS lifetime was determined from the slope of the first-order plots to be 670 μs.[85]

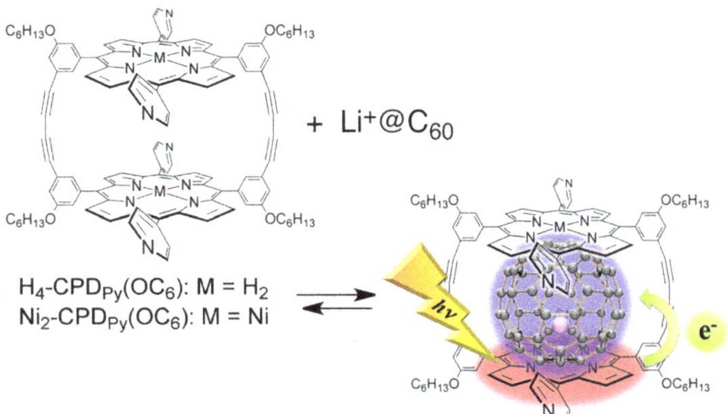

Figure 3.19 Supramolecular complex formation and photoinduced charge separation between $MCPD_{Py}(OC_6)$ and $Li^+@C_{60}$. Reproduced from ref. 85 with permission from the Royal Society of Chemistry.

The CS state was also observed for $Li^+@C_{60}\subset H_4$–$CPD_{Py}(OC_6)$. The quantum yields of the CS states were estimated to be 13% for $Li^+@C_{60}\subset Ni_2$–$CPD_{Py}(OC_6)$ and 32% for $Li^+@C_{60}\subset H_4$–$CPD_{Py}(OC_6)$.[85] When $Li^+@C_{60}$ was replaced by pristine C_{60}, no CS states were produced as predicted by their higher energy levels than those of the triplet excited states of $CPD_{Py}(OC_6)$ and C_{60}.[86]

The mechanisms of photoinduced charge separation in the $Li^+@C_{60}\subset Ni_2$–$CPD_{Py}(OC_6)$ supramolecular complex are shown in Scheme 3.8.[85] Laser photoexcitation of $Li^+@C_{60}\subset Ni_2$–$CPD_{Py}(OC_6)$ at 420 nm resulted in the generation of a singlet excited state of Ni_2–$CPD_{Py}(OC_6)$ ($^1[H_4\text{-}CPD_{Py}(OC_6)]^*$), followed by intersystem crossing to produce $^3[Ni_2$–$CPD_{Py}(OC_6)]^*$.[85] Then, ET from $^3[Ni_2$–$CPD_{Py}(OC_6)]^*$ to $Li^+@C_{60}$ occurs with the driving force of 0.44 eV to produce the CS state. The CS state decays slowly *via* intrasupramolecular BET with the lifetime of 670 μs (Scheme 3.8).[85]

When porphyrin dimers were replaced by a porphyrin trimer ($TPZn_3$), a fullerene derivative containing a pyridine moiety (PyC_{60}) was inserted inside the cavity of the tripod conformation of $TPZn_3$ (Figure 3.20) by π–π interactions combined with the coordination of the pyridine moiety to Zn^{2+} (Scheme 3.9).[87] The formation constant of the $TPZn_3$–PyC_{60} complex (1.1×10^5 M^{-1}) in toluene at 298 K determined from the UV-vis absorption spectral titration is much larger than those of the corresponding monomer (MPZn) and dimer porphyrin ($DPZn_2$) because of the stronger π–π interaction inside the cavity.[87]

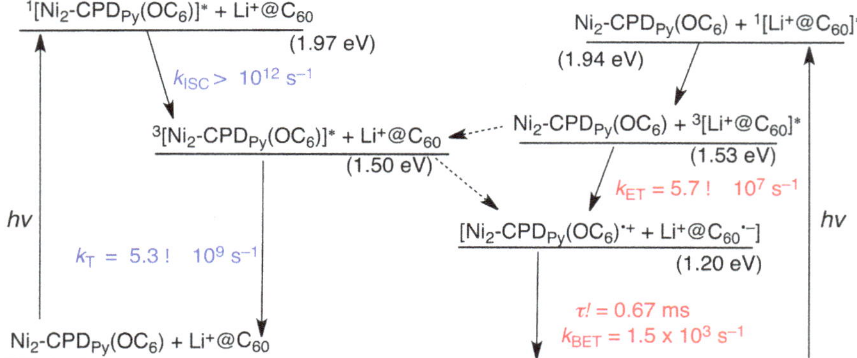

Scheme 3.8 Energy diagram for photoinduced charge separation and charge recombination of $Li^+@C_{60}\subset Ni_2$-$CPD_{Py}(OC_6)$; broken arrow: minor pathway. Reproduced from ref. 85 with permission from the Royal Society of Chemistry.

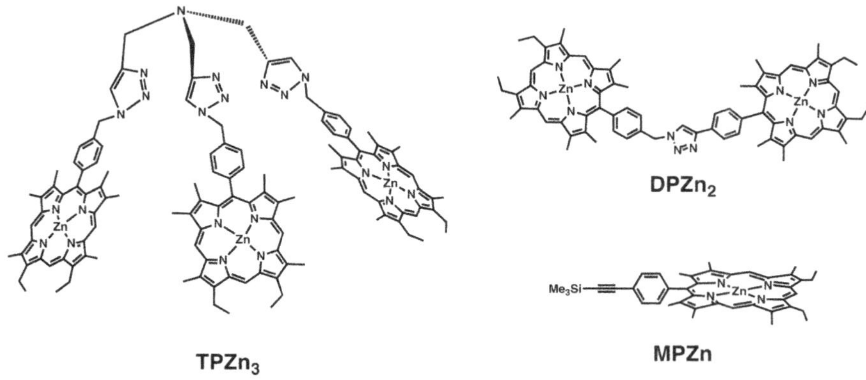

Figure 3.20 A porphyrin tripod and the reference dimer and monomer.

$$TPZn_3 + PyC_{60} \xrightleftharpoons{K} (TPZn_3\text{–}PyC_{60})$$

PyC60

Scheme 3.9 Formation of a supramolecular complex between $TPZn_3$ and PyC_{60}.

The encapsulation of PyC_{60} inside the cavity of $TPZn_3$ was confirmed by the ^1H NMR signals of $TPZn_3$, which exhibited downfield shifts upon complexation with PyC_{60}, whereas the pyridyl protons of PyC_{60} exhibited large upfield shifts by the complexation due to the large porphyrin aromatic ring current.[87] The encapsulation of PyC_{60} inside the cavity of $TPZn_3$ was also supported by the DFT-optimized structure (B3LYP/3-21G(*) basis set).[87]

Photoexcitation of the $TPZn_3$–PyC_{60} complex resulted in ET from the singlet excited state of $TPZn_3$ to PyC_{60} to produce the CS state ($TPZn_3^{\bullet +}$–$PyC_{60}^{\bullet -}$) as shown by femtosecond laser-induced transient absorption measurements, where the decay of the absorption due to $^1TPZn_3^*$ was accompanied by appearance of the absorption bands at $\lambda_{max} = 1000$ nm due the $PyC_{60}^{\bullet -}$ and 670 nm due to $TPZn_3$ radical cation ($TPZn_3^{\bullet +}$).[87]

In sharp contrast to the case of the $TPZn_3$–PyC_{60} complex, the transient absorption spectrum of monomer porphyrin (MPZn) in the presence of PyC_{60} exhibited no formation of the CS state.[87] Only energy transfer from $^1MPZn^*$ to PyC_{60} occurred to afford the singlet excited state $^1PyC_{60}^*$ (1.76 eV), followed by ISC to $^3PyC_{60}^*$ at 2800 ps, accompanied by the recovery of the ground state, PyC_{60}.[87]

The energy diagrams of photoinduced ET in the $TPZn_3$–C_{60} complex and MPZn in the presence of PyC_{60} in toluene are shown in Scheme 3.10a and b, respectively.[87] The energy level (1.49 eV) of the CS state ($TPZn_3^{•+}$–$PyC_{60}^{•-}$) is lower than that of the triplet excited state of the PyC_{60} moiety (1.56 eV).[87] The rate constant (k_{ET}) of photoinduced ET from $^1TPZn_3^*$ to PyC_{60} is larger than that of intersystem crossing.[87] The CS lifetime was determined to be $\tau_{CS} = 0.53$ ns from the rate constant of BET ($k_{BET} = 1.9 \times 10^9$ s^{-1}).[87] In contrast to the $TPZn_3$–C_{60} complex, MPZn forms no complex with PyC_{60}, when only energy transfer from $^1MPZn^*$ to PyC_{60} occurred to produce $^1PyC_{60}^*$, followed by ISC to afford $^3PyC_{60}^*$.[87]

$TPZn_3$ also forms a supramolecular 1 : 1 complex with gold(III)tetra (4-pyridyl)porphyrin ($AuTPyP^+$) by encapsulation of $AuTPyP^+$ inside the cavity of $TPZn_3$ *via* triple coordination bonds in nonpolar solvents.[88] Photoexcitation of the $TPZn_3$–$AuTpyP^+$ also resulted in

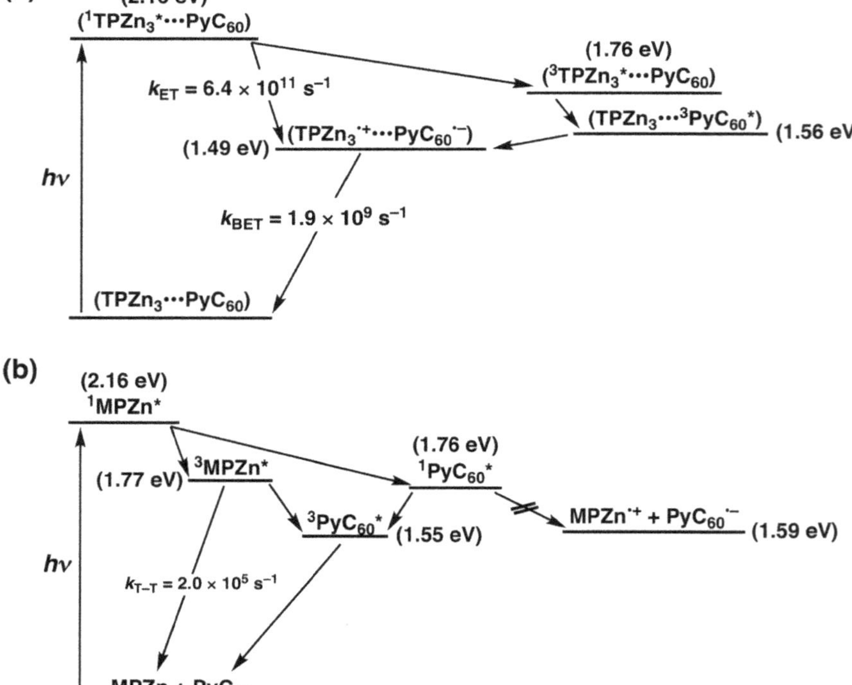

Scheme 3.10 Photoinduced ET and BET of (a) $TPZn_3$ and (b) MPZn in the presence of PyC_{60} in toluene. Reproduced from ref. 87 with permission from American Chemical Society, Copyright 2010.

efficient ET from $^1TPZn_3{}^*$ to $AuTPyP^+$ to produce the ET state ($TPZn_3{}^{\bullet+}$–AuTpyP) as in the case of the $TPZn_3$–PyC_{60} complex.[88]

Mimicking light harvesting and CS processes in photosynthesis has been achieved by using zinc porphyrin oligomers [$P(ZnP)_8$] with oligopeptidic backbones in which porphyrins are held in a favoured spacing and orientation by fairly short helical oligopeptides to accommodate fullerenes (PyC_{60} and ImC_{60}) between porphyrins by π–π interaction (Figure 3.21).[89] $P(ZnP)_8$ forms a strong supramolecular complex with PyC_{60} by the strong π–π interaction between two zinc porphyrins and $C_{60}Py$ in addition to the axial coordination of the pyridine moiety of $C_{60}Py$ to the zinc porphyrin.[89] ImC_{60} bearing an imidazole coordinating ligand is more strongly bound by the oligopeptide [$P(ZnP)_8$] than PyC_{60}.[89] Efficient energy transfer between the ZnP molecules occurred in the supramolecular complex of [$P(ZnP)_8$] with ImC_{60} judging from the much larger fluorescence quenching constant (3.3×10^5 M^{-1}) as compared with the formation constant of the ground state (1.5×10^4 M^{-1}).[89]

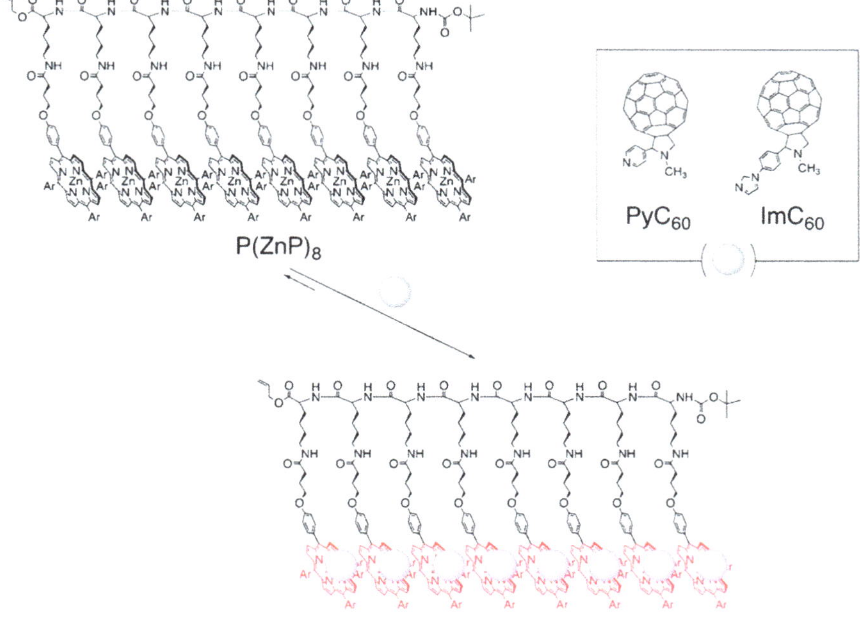

Figure 3.21 Formation of a supramolecular complex: the porphyrin–peptide octamer [$P(ZnP)_8$] with PyC_{60} or ImC_{60}. Ar = 3,5-(*t*-Bu)$_2C_6H_3$. Reproduced from ref. 89, https://doi.org/10.1149/2162-8777/abaaf5, under the terms of the CC BY 4.0 license, https://creativecommons.org/licenses/by/4.0/.

The laser photoexcitation at 561 nm of the supramolecular complex of $P(ZnP)_8$ with PyC_{60} resulted in the formation of the CS state $[P(ZnP)_8^{\bullet+}–PyC_{60}^{\bullet-}]$ as shown by the transient absorption spectra, where the absorption band due to $PyC_{60}^{\bullet-}$ was observed at 1000 nm together with that due to $ZnP^{\bullet+}$ at 630 nm upon laser excitation of the ZnP moiety.[89] The decay of the CS state obeyed first-order kinetics, and the first-order plots with different laser powers afforded linear correlations with the same slope, indicating that the CS state decayed *via* intracomplex BET in the supramolecular complex rather than intermolecular BET from $PyC_{60}^{\bullet-}$ to $P(ZnP)_8^{\bullet+}$ in different supramolecular complexes.[89] It was confirmed that the second-order plots gave no linear correlations.[89] The lifetime of the CS state $[P(ZnP)_8^{\bullet+}–ImC_{60}^{\bullet-}]$ was determined to be 840 μs, which is a bit longer than that of $[P(ZnP)_8^{\bullet+}–PyC_{60}^{\bullet-}]$ (600 μs) in PhCN at 298 K.[89] Such efficient energy transfer among ZnP molecules in $P(ZnP)_8$, followed by rapid charge separation to afford the long lived CS state, mimics well light-harvesting and charge-separation processes in photosynthesis.[89]

The π–π interaction also plays a crucial role in formation of supramolecular nanohybrids between porphyrin–peptide hexadecamer $[P(H_2P)_{16}]$[90] that wraps with single walled carbon nanotube (SWNTs),[91] where π–π interaction between porphyrins and nanotubes resulted in the extraction of large-diameter nanotubes (*ca.* 1.3 nm), as indicated by UV-vis-near IR and Raman spectroscopic measurements as well as TEM images.[91] Laser pulse excitation of $P(H_2P)_{16}$/SWNTs in DMF resulted in ET from the singlet excited states as well as the triplet excited states of $P(H_2P)_{16}$ to SWNTs to produce the CS state, which exhibited transient absorption bands at 450 nm together with a broad absorption band in the 500–800 nm region due to a free base porphyrin radical cation and a number of bleaching bands due to the porphyrin (518, 553, 597, and 649 nm) and SWNTs (750 nm).[91] The CS lifetime was determined to be 370 ± 30 μs from the first-order decay of the absorbance due to the CS state.[91]

Porphyrin molecules can also be assembled by using porphyrin dendrimers in which the morphology and the photochemical function of porphyrin dendrimers are similar to those of the light-harvesting units in photosynthesis and they can be combined with photosynthetic reaction centre units. For example, a zinc porphyrin dendrimer $[D(ZnP)_{16}]$[92] forms a supramolecular complex with *N*-methyl-2-(4′-pyridyl)-3,4-fulleropyrrolidine (PyC_{60}),[93] which has a pyridine binding site, as shown in Figure 3.22.[94] The porphyrin dendrimers were synthesized by coupling of the porphyrin activated ester [5-amino-2-{5,10,15,20-tetrakis(3,5-di-*tert*-butylphenyl)}]-5-oxopentanoic

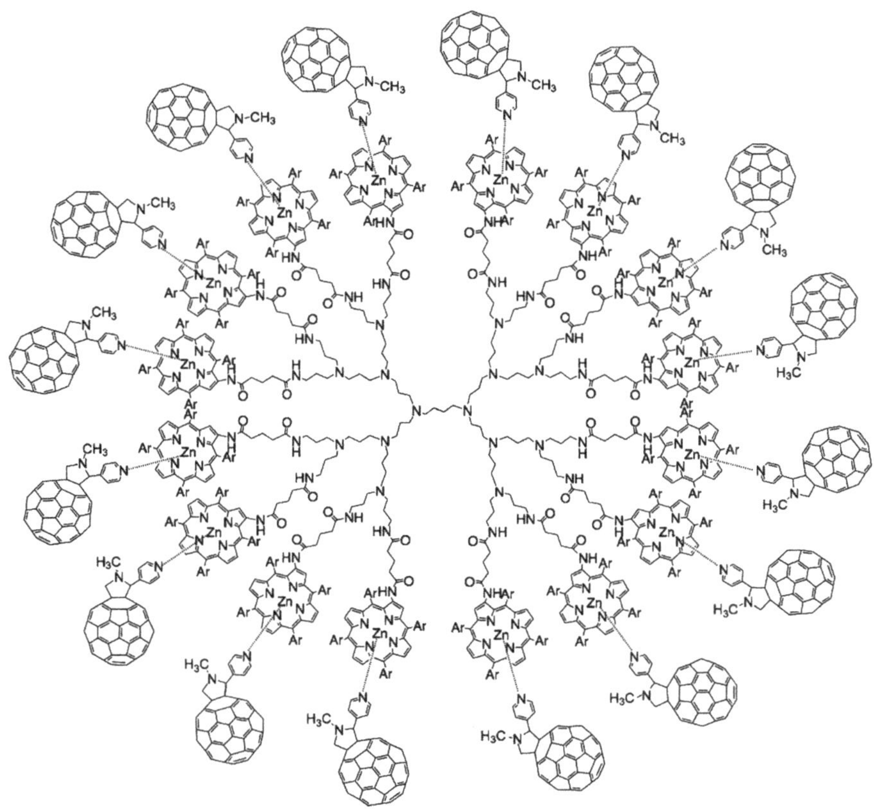

Figure 3.22 A supramolecular complex between a zinc porphyrin dendrimer [D(ZnP)$_{16}$] and PyC$_{60}$. Reproduced from ref. 94 with permission from the Royal Society of Chemistry.

acid 2,5-dioxopyrrolidin-1-yl ester with the first, second, or third generation polypropylene-imine dendrimer.[94]

Photoexcitation of the Soret band of D(ZnP)$_{16}$ at 438 nm in PhCN resulted in fluorescence emission at $\lambda_{max} = 609$ and 645 nm.[94] The fluorescence was efficiently quenched by addition of PyC$_{60}$ to a PhCN solution of D(ZnP)$_{16}$ due to the binding of PyC$_{60}$ to D(ZnP)$_{16}$.[94] The binding constant (K) of the 1 : 1 ZnP monomer unit–PyC$_{60}$ complex was determined from the fluorescence quenching to be 5.0×10^4 M^{-1}, which is significantly larger than the K value (1.6×10^4 M^{-1}) determined from the UV-vis absorption spectral change.[94] The larger K value determined from the fluorescence quenching than the value from the absorption spectral change results from efficient energy transfer between the porphyrin units, because the fluorescence of the ZnP moiety, to which no PyC$_{60}$ is bound, is quenched by PyC$_{60}$ bound to a different ZnP moiety *via* the efficient energy transfer between the adjacent ZnP moieties.[94] In contrast to the case of the case $D(ZnP)_{16}$, no fluorescence quenching of a ZnP monomer by PyC_{60} occurred, because binding of PyC_{60} to a ZnP monomer was too weak to compete with the coordination of the solvent (PhCN).[94]

Photoexcitation of the supramolecular complex of $D(ZnP)_{16}$ with PyC_{60} at 561 nm, where only the ZnP moiety was excited, resulted in efficient energy migration, followed by ET from the singlet excited state of ZnP to the PyC_{60} bound to ZnP to produce the CS state, as revealed by nanosecond laser transient absorption measurements.[94] The absorption band due to $PyC_{60}^{\bullet -}$ in the CS state was observed at 1000 nm together with that due to $ZnP^{\bullet +}$ at 630 nm.[94] The quantum yield (Φ) of formation of the CS state of the supramolecular complex of $D(ZnP)_{16}$ with PyC_{60} was determined to be 25%.[94] The decay of the absorption band at 1000 nm due to $PyC_{60}^{\bullet -}$ obeyed first-order kinetics with the same slope irrespective of different laser power intensities to afford the CS lifetime of 250 μs in PhCN at 298 K.[94]

3.3.2 Cation–Anion Binding

Electrostatic interactions between cation and anion species are strong enough to form stable supramolecular complexes between an electron donor and an acceptor by cation–anion binding. For example, anionic porphyrin sulfonates (H_2TPPS^{4-} and $ZnTPPS^{4-}$) that act as electron donors at the photoexcited states form strong supramolecular complexes with a cationic lithium ion-encapsulated fullerene ($Li^+@C_{60}$)[95] that acts as an electron acceptor.[96] By addition of a PhCN solution of $Li^+@C_{60}$, the Soret band of freebase tetraphenylporphyrin tetrasulfonate $[(Bu_4N^+)_4H_2TPPS^{4-}]$ decreased significantly and the absorption maximum was red-shifted to 427 nm with an isosbestic point at 430 nm in PhCN at 298 K.[96] The absorbance change exhibited saturation behaviour to approach a constant value with increasing concentration of $Li^+@C_{60}$.[96] The formation constant (K) of the $H_2TPPS^{4-}/Li^+@C_{60}$ complex was determined to be 3.0×10^5 M^{-1} in PhCN at 298 K.[96] A Job plot indicates that the binding stoichiometry was 1 : 1.[96] When H_2TPPS^{4-} was replaced by $ZnTPPS^{4-}$, the K value was determined to be 1.6×10^5 M^{-1} in PhCN at 298 K.[96] The same K values were obtained from fluorescence quenching of $^1H_2TPPS^{4-*}$ by $Li^+@C_{60}$ in PhCN at 298 K.[96] Thus, strong supramolecular complexes $[(ZnTPPS^{4-})–Li^+@C_{60}$ and $(H_2TPPS^{4-})–Li^+@C_{60}]$ are produced by anion–cation binding as well as π–π interactions.[96]

The fluorescence quenching of $^1H_2TPPS^{4-*}$ and $^1ZnTPPS^{4-*}$ by $Li^+@C_{60}$ occurred *via* energy transfer from $^1H_2TPPS^{4-*}$ and $^1ZnTPPS^{4-*}$ to $Li^+@C_{60}$, as revealed by the femtosecond laser transient absorption measurements.[96] Nanosecond laser excitation of $(ZnTPPS^{4-})$–$Li^+@C_{60}$ and (H_2TPPS^{4-})–$Li^+@C_{60}$ resulted in ET from H_2TPPS^{4-} and $ZnTPPS^{4-}$ to $^3Li^+@C_{60}{}^*$ to produce the CS state, exhibiting the absorption band of $(H_2TPPS^{4-})^{\bullet+}$ ($\lambda_{max} = 670$ nm)[97] together with that of $Li^+@C_{60}{}^{\bullet-}$ ($\lambda_{max} = 1035$ nm).[96] The decay of the CS state $[(H_2TPPS^{4-})^{\bullet+}$–$Li^+@C_{60}{}^{\bullet-}]$ in PhCN obeyed first-order kinetics with use of different laser powers.[96] The first-order plots afforded the same slope irrespective of the difference in concentration of the CS state, indicating that there was no or little contribution from bimolecular BET from $Li^+@C_{60}{}^{\bullet-}$ to $[H_2TPPS^{4-}]^{\bullet+}$ in different supramolecular complexes.[96] The lifetimes of the triplet CS states of the supramolecular complexes were determined to be 310 μs for H_2TPPS^{4-} and 300 μs for $ZnTPPS^{4-}$ in PhCN at 298 K.[96] These are the longest lifetimes of the CS states ever reported for monomer porphyrin/fullerene dyads linked by noncovalent bonds in solution.[96] The quantum yield of the CS state was determined to be 39% using the absorption of the CS state ($Li^+@C_{60}{}^{\bullet-}$: $\varepsilon_{1035} = 7300$ M^{-1} cm^{-1}).[96]

The temperature dependence of BET in the supramolecular complexes afforded $\Delta H^{\neq} = 3.0$ kcal mol^{-1} for $(ZnTPPS^{4-})^{\bullet+}$–$Li^+@C_{60}{}^{\bullet-}$ and 5.4 kcal mol^{-1} for $(H_2TPPS^{4-})^{\bullet+}$–$Li^+@C_{60}{}^{\bullet-}$.[96] This indicates that there is a significant energy difference between the singlet and triplet CS states and that the BET in the supramolecular complexes may proceed *via* the thermally activated singlet CS state.[96] The extrapolation of the Eyring plot predicts the lifetime of the CS state $[(H_2TPPS^{4-})^{\bullet+}$–$Li^+@C_{60}{}^{\bullet-}]$ at 77 K to be 60 h.[96] Such a long-lived triplet CS state was indeed detected by the EPR measurements by photoirradiation of the $(ZnTPPS^{4-})$–$Li^+@C_{60}$ complex and the H_2TPPS^{4-}–$Li^+@C_{60}$ complex in frozen PhCN.[96] The spin–spin interaction in the triplet CS state of the supramolecular complex is indicated by the fine structure observed at 77 K.[96] From the zero-field splitting values ($D = 52$ G for $ZnTPPS^{4-}$ and 56 G for H_2TPPS^{4-}), the distances (r) between two electron spins in the $(ZnTPPS^{4-})^{\bullet+}$–$Li^+@C_{60}{}^{\bullet-}$ complex and the $(H_2TPPS^{4-})^{\bullet+}$–$Li^+@C_{60}{}^{\bullet-}$ complex were estimated by the relation $D = 27\,800/r^3$ to be 8.1 and 7.9 Å, respectively.[96] These r values agree with the centre-to-centre distance of a reported crystal structure of porphyrin–C_{60} complexes.[96]

Supramolecular electron donor–acceptor complexes are also constructed by using anion–cation binding between one tetra-anionic porphyrin ($1\text{-}M^{4-}$: M = H_2 and Zn) and two dicationic porphyrins ($2\text{-}Zn^{2+}$), which were produced by the two-electron oxidation of a π-extended 1,3-dithiol-2-ylidene quinoidal porphyrin that is a porphyrin-bridged TTF (tetrathiafulvalene) in Figure 3.23.[98] The X-ray crystal structure of a 2:1 supramolecular complex between two $2\text{-}Zn^{2+}$ and $1\text{-}M^{4-}$ (sandwich type) showed that both the top and bottom faces were covered by large charged molecules.[98] The formation constants of $1\text{-}Zn^{4-}/(2\text{-}Zn^{2+})_2$ and $1\text{-}H_2{}^{4-}/(2\text{-}Zn^{2+})_2$ in PhCN at 298 K were determined by the visible absorption spectral titrations to be $5.9(5) \times 10^{11}$ and $6.6(5) \times 10^{12}$ M^{-2}, respectively.[98] The large formation constants results from the strong coulombic (electrostatic) interaction between the two dicationic porphyrin and the tetraanionic porphyrin.[98]

Photoexcitation of the $1\text{-}H_2{}^{4-}/(2\text{-}Zn^{2+})_2$ complex resulted in ET from $1\text{-}M^{4-}$ to $2\text{-}Zn^{2+}$ *via* the triplet excited state to afford the triplet CS state as detected by transient absorption spectra.[98] The decay of the triplet CS states obeyed first-order kinetics to afford the same slope irrespective of differences in the initial concentration of the CS state, indicating that intracomplex BET proceeds in the supramolecular complex and there is no contribution from intermolecular BET between two supramolecular complexes.[98] The intermolecular BET processes are prohibited, because the radical trianionic porphyrin $1\text{-}M^{\bullet 3-}$ in the CS state is sandwiched by two positively charged moieties, $2\text{-}Zn^{2+}$ and $2\text{-}Zn^{\bullet +}$, to prevent close contacts with another supramolecular complex.[98] The lifetimes of the CS states of $1\text{-}H_2{}^{\bullet 3-}/(2\text{-}Zn^{\bullet +})(2\text{-}Zn^{2+})$ and $1\text{-}Zn^{\bullet 3-}/(2\text{-}Zn^{\bullet +})(2\text{-}Zn^{2+})$ in PhCN at 298 K were determined to be 83 ms and 43 ms, respectively. These are the longest lifetimes ever reported among the CS states of supramolecular donor–acceptor complexes in solution.[98] The temperature dependence of intramolecular BET in the CS state of the supramolecular complex $[1\text{-}H_2{}^{\bullet 3-}/(2\text{-}Zn^{\bullet +})(2\text{-}Zn^{2+})]$ was analysed based on the Marcus equation [eqn (2.20)] to afford values of λ and V of 0.24 eV and 0.16 cm^{-1}, respectively.[98] The small V value results from the spin-forbidden BET of the triplet CS state to the singlet ground state and small orbital interactions between $1\text{-}H_2{}^{\bullet 3-}$ and $2\text{-}Zn^{\bullet +}$ due to the slipped-sandwich arrangement (Figure 3.23).[98] The small λ value may result from the small bond reorganization energy of the porphyrin moieties with the large 2D π-plane and small solvent reorganization energy due to the limited space between the electron donor and acceptor moieties.[98]

Cation–anion binding has also been utilized to form a supramolecular complex between diprotonated [30]octaphyrin(0.0.0.0.0.0.0.0)

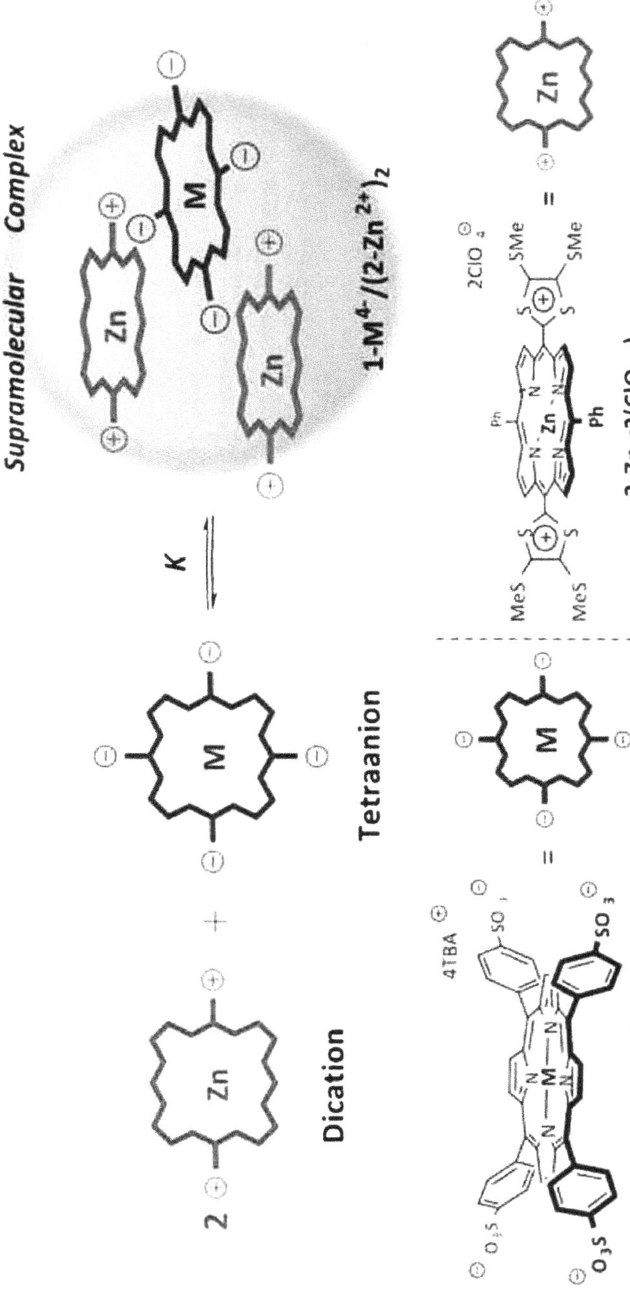

Figure 3.23 Formation of supramolecular porphyrin complexes between one tetraanionic porphyrin and two dicationic porphyrins, 1-M^{4-}/(2-Zn^{2+})$_2$; M = H_2 or Zn, by cation–anion binding. Reproduced from ref. 98 with permission from the Royal Society of Chemistry.

(cyclo[8]pyrrole; C8) and pyrenecarboxylate anion as the TBA (tetrabutylammonium) salt (Py) (Figure 3.24a).[99] The formation constant (K_a) of the C_8–Py complex was determined by the visible absorption spectral titration (Figure 3.24b) to be $(2.6 \pm 0.3) \times 10^5$ M^{-1} in MeCN at 298 K.[99] Photoexcitation of a deaerated MeCN solution of the C_8–Py complex resulted in ET from the C_8 moiety to the singlet excited state of the Py moiety ($^1Py^*$) to produce the CS state ($C_8^{\bullet+}$–$Py^{\bullet-}$), which has absorption bands due to $C_8^{\bullet+}$ (740 and 820 nm) and $Py^{\bullet-}$ (480 nm) rather than those due to $C_8^{\bullet-}$ and $Py^{\bullet+}$.[99] It is interesting to note that the CS energy of $C_8^{\bullet+}$–$Py^{\bullet-}$ (2.58 eV) is much higher than the reversed CS energy of $C_8^{\bullet-}$–$Py^{\bullet+}$ (1.31 eV).[99] The CS process to produce the lower energy CS state ($C8^{\bullet-}$–$Py^{\bullet+}$) has the large driving force of

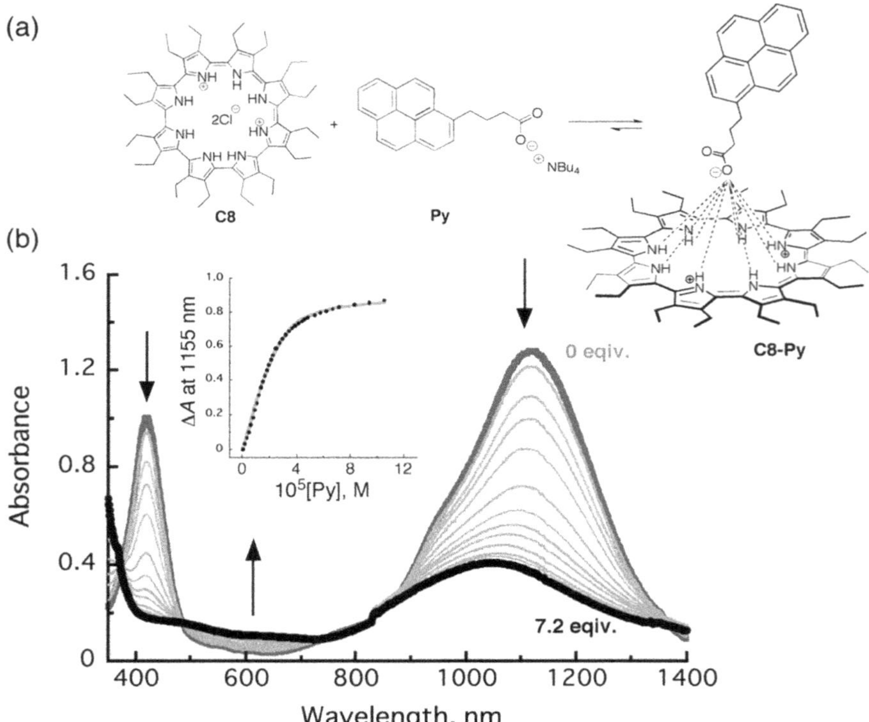

Figure 3.24 (a) Formation of a supramolecular complex between cyclo[8]pyrrole (C_8) and tetra-*n*-butylammonium 1-pyrenebutyrate (Py) by cation–anion binding. (b) UV-vis absorption spectral titration of Py added into an MeCN solution of C8 (1.5×10^{-5} M) at 298 K. Inset: Curve fit (line) to a 1:1 binding isotherm produced from the absorbance spectral change at 1155 nm (points). Reproduced from ref. 99 with permission from American Chemical Society, Copyright 2008.

2.15 eV, which is deep in the Marcus inverted region, where the rate constant of ET from the C_8 moiety to the $^1Py^*$ moiety may be much smaller than that of ET from the $^1Py^*$ moiety to the C_8 moiety to produce the high energy CS state of $C_8^{\bullet +}$–$Py^{\bullet -}$ in the Marcus normal region with the driving force of 0.88 eV.[99]

The charge reversal seen in the 'umpolung' system in the C_8–Py complex has also been reported for ET from electron donors to $^1Py^*$.[100] The CS rate constant was determined from the fluorescence lifetime measurements of $^1Py^*$ in the presence of C_8 to be 5.1×10^6 s^{-1} in MeCN at 298 K.[99] The decay of the CS state obeyed first-order kinetics and the CS lifetime was determined to be as long as 300 μs in MeCN at 298 K. The CS state decayed to the triplet excited state ($^3C_8^*$) rather than to the ground state (C_8) as confirmed by the transient absorption measurements of the C_8–Py complex recorded at 900 μs, which agreed with that due to $^3C_8^*$ produced upon photoexcitation of C_8 alone.[99] The BET to produce $^3C_8^*$ rather than the ground state C_8 is also understood in terms of the Marcus theory of ET, because the BET to the ground state has a much larger driving force located deep in the Marcus inverted region to exhibit a much slower rate than that of BET to the triplet excited-state in the Marcus normal region.[99] It was confirmed that the use of pyrene-1-butyric acid in conjunction with C_8 under otherwise identical conditions afforded none of the peaks characteristic of $C_8^{\bullet +}$ upon photoexcitation.[99] Thus, the anion binding is essential for formation of the supramolecular complex.[99]

Cation–anion binding also plays an important role in the formation of a 1 : 2 supramolecular complex [$H_4DPP(FcCOO)_2$] between diprotonated dodecaphenylporphyrin (H_4DPP^{2+}) and a ferrocene carboxylate anion ($FcCOO^-$) as indicated by the UV-vis spectral titration in PhCN.[101] A Job's plot confirmed the 1 : 2 stoichiometry of the complex formation. The formation constant of $H_4DPP(FcCOO)_2$ was determined from the UV-vis absorption spectral change to be 9.2×10^8 M^{-2} in MeCN at 298 K.[101] The formation of the hydrogen bonds was supported by the ^1H NMR spectral changes observed by the addition of FcCOOH, which resulted in significant change in the chemical shifts due to the formation of hydrogen bonds between the pyrrole NH protons of H_4DPP^{2+} and the carboxylate group of $FcCOO^-$.[101]

The X-ray crystal structure of the H_4DPP–$(FcCOO)_2$ complex is shown in Figure 3.25, where two ferrocene-carboxylate ($FcCOO^-$) anions bind with H_4DPP^{2+} from the top and bottom sides by cation–anion electrostatic interaction as well as hydrogen bonds in an

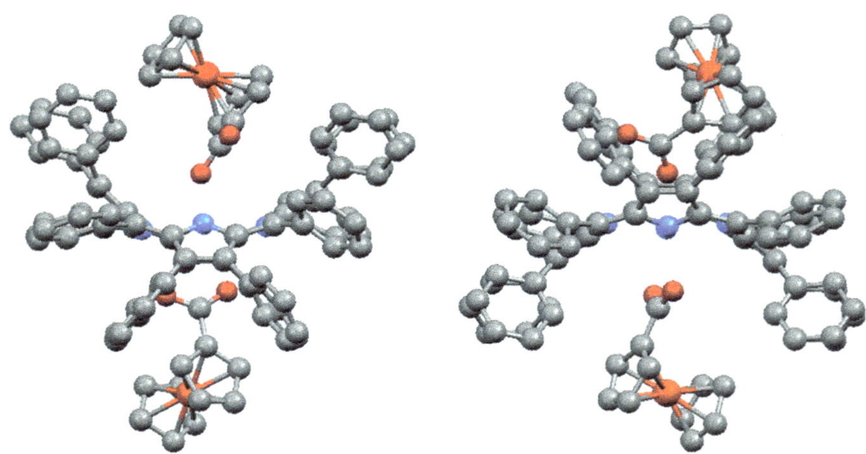

Figure 3.25 X-ray crystal structure of the supramolecular $H_4DPP–(FcCOO)_2$ complex seen from different directions. Reproduced from ref. 101 with permission from American Chemical Society, Copyright 2010.

asymmetric fashion.[101] On one side of the saddle-distorted H_4DPP^{2+} macrocycle, two-point hydrogen bonds are formed between two of the pyrrole N–H protons and two of the oxygen atoms in $FcCOO^-$.[101] On the other side of the ring, hydrogen bond interactions are also observed between two of the four N–H pyrrolic protons and one of the $FcCOO^-$ carboxylate oxygen atom.[101] Both $FcCOO^-$ anions reside in a cleft-like space made by the two phenyl groups attached to the pyrroles of the porphyrins (Figure 3.25).[101]

The HOMO and LUMO orbitals of the $H_4DPP–(FcCOO)_2$ complex are localized on the $FcCOO^-$ moiety and the H_4DPP^{2+} moiety, respectively.[101] Thus, photoinduced ET occurs from the $FcCOO^-$ moiety to the excited state of H_4DPP^{2+}.[101] Rate constants of photoinduced ET and BET in the $H_4DPP(FcCOO)_2$ complex were determined by femtosecond laser transient absorption measurements.[101] Femtosecond laser excitation at 430 nm of a deaerated PhCN solution of $H_4DPP(FcCOO)_2$ resulted in ET from the $FcCOO^-$ moiety to the singlet excited state of the H_4DPP^{2+} moiety ($^1[H_4DPP^{2+}]^*$) to produce the ET state.[101] The rate constant of ET from $^1[H_4DPP^{2+}]^*$ to FcCOO were determined to be 5.0×10^{11} s^{-1} and 6.1×10^{10} s^{-1}, respectively.[101] Similarly, the rate constants of ET and BET were determined using the supramolecular complexes of H_4DPP^{2+} with other ferrocene carboxylate derivatives, which were formed using the anion linking motif.[101] The driving force dependence of $\log k_{ET}$ and $\log k_{BET}$ is shown in Figure 3.26, where the fitting solid line was drawn based on

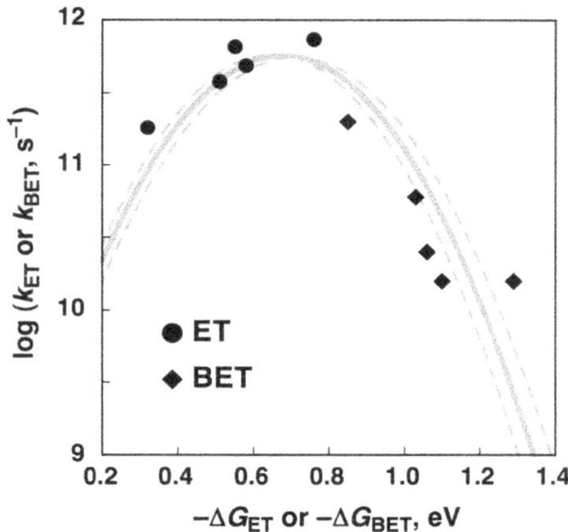

Figure 3.26 Driving force dependence of log k_{ET} (●) or k_{BET} (◆) for intrasupramolecular ET from $DCOO^-$ to $^1[H_4DPP^{2+}]^*$ and BET from $H_4DPP^{\bullet+}$ to D^+COO^- in supramolecular complexes of H_4DPP^{2+} with ferrocene carboxylate derivatives ($DCOO^-$) in PhCN at room temperature. Fitting to the Marcus equation for nonadiabatic ET [eqn (2.20)] is shown by the solid line with use of $\lambda = 0.68$ eV and $V = 43$ cm^{-1} and the dotted lines with $\lambda = 0.65$ and 0.71 eV and $V = 43$ cm^{-1}, respectively. Reproduced from ref. 101 with permission from American Chemical Society, Copyright 2010.

the Marcus theory of nonadiabatic ET [eqn (2.20)] to afford a small reorganization energy ($\lambda = 0.68(3)$ eV) and a large electronic coupling matrix element ($V = 43(7)$ cm^{-1}).[101] The λ value is comparable to those of covalently linked donor–acceptor systems composed of porphyrins as electron donors (0.41–0.66 eV).[28,30] The small λ value made it possible for BET processes with a large BET driving force to be located deep in the Marcus inverted region ($-\Delta G_{BET} \gg \lambda$), where the CS lifetime becomes longer with increasing BET driving force (Figure 3.26).[101] The large V value may result from the strong cation–anion binding as well as hydrogen-bonding interactions between the electron donor and acceptor moieties in the supramolecular complex.[101] The energy level of $^1[H_4DPP^{2+}]^*$ is higher than that of the ET state, $[H_4DPP^{\bullet+}](D^+COO)(DCOO^-)]$.[101]

The dependence of ln k_{ET} on distance between the electron donor (D) and acceptor (A) moieties in the supramolecular complexes was also examined by using ferrocene carboxylic acid derivatives bearing linear phenylene linker(s) between the ferrocene moiety and the carboxyl group. The distance (r) between the D and A moieties is defined as that between the Fe atom of ferrocene and the centre of the mean plane of the porphyrin ring in H_4DPP^{2+}.[101] The dependence of the rate constant for intrasupramolecular ET (k_{ET}) on the distance between the D and A moieties is expressed by eqn (3.5),

$$\ln k_{ET} = \ln k_0 - \beta r \tag{3.5}$$

where k_0 is the rate constant for adiabatic intrasupramolecular ET, being independent of r, which is the D–A centre-to-centre distance, and β is the decay coefficient factor (damping factor) that depends primarily on the bonding nature of the bridging molecule.[102] A linear correlation between $\ln k_{ET}$ and r affords the β value of 0.64 Å^{-1},[101] which is similar to the value (0.60 Å^{-1}) for D–A linked molecules linked with various r values using covalent bonds such as ZnP–C_{60}, Fc–ZnP–C_{60}, Fc–H_2P–C_{60}, ZnP–H_2P–C_{60} and Fc–ZnP–H_2P–C_{60}.[22] Thus, the D–A linkage by cation–anion binding with hydrogen bonding in supramolecular complexes provides the electronic coupling matrix element, which is large enough to assure efficient ET through the noncovalent bonding as well as covalent bonding.[101]

3.4 Binding of Metal Ions

A biomimetic model of the bacterial 'special pair' donor, a cofacial zinc phthalocyanine dimer, was made *via* K^+ ion induced dimerization of 4,5,4′,5′,4″,5″,4‴,5‴-zinc tetrakis(1,4,7,10,13-pentaoxatridecamethylene)phthalocyanine.[103] The dimer forms a supramolecular complex with functionalized fullerenes *via* 'two-point' binding involving axial coordination and crown ether-alkyl ammonium cation complexation to form the D–A supramolecular complex $[K_4[ZnTCPc]_2{:}(pyC_{60}NH_3{}^+)_2]$ mimicking the noncovalently bound entities of the bacterial photosynthetic reaction centre, as shown in Figure 3.27, where the chemical and DFT optimized structures reveal how the D and A moieties are assembled.[103]

Laser photoexcitation at 410 nm of a deaerated PhCN solution of the supramolecular $K_4[ZnTCPc]_2{:}(pyC_{60}NH_3{}^+)_2$ complex resulted in formation of the CS state, which showed transient absorption bands at 870 and 1000 nm due to $ZnTCPc^{\bullet+}$ and $C_{60}{}^{\bullet-}$, respectively.[103] The quantum yield of the CS state was determined to be $29 \pm 6\%$.[103] The decay of the absorption at 870 and 1000 nm due to the CS state obeyed first-order kinetics to afford a CS lifetime of 6.7 μs.[103] Thus, binding of K^+ ions to the four crown ether units of zinc phthalocyanine provided a scaffold to form the supramolecular $K_4[ZnTCPc]_2{:}(pyC_{60}NH_3{}^+)_2$ complex, which afforded the long CS lifetime upon photoexcitation.[103]

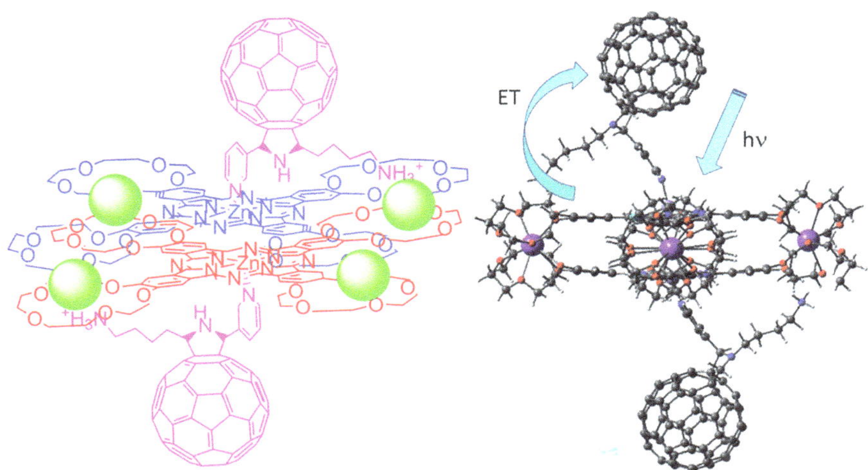

Figure 3.27 Chemical structure and B3LYP/3-21G(*) optimized structure of the supramolecular $K_4[ZnTCPc]_2{:}(pyC_{60}NH_3^+)_2$ complex and photoinduced ET. Reproduced from ref. 103 with permission from American Chemical Society, Copyright 2009.

Binding of metal ions to the CS state has also been reported to affect the energy levels and the lifetimes of the CS state of D–A linked molecules.[104] Notable effects of metal ions on the rate constants of photoinduced ET and BET in a porphyrin-containing donor–acceptor linked molecule, a zinc porphyrin–naphthalenediimide (ZnP–NIm) dyad, have been reported as shown in Scheme 3.11.[105] The photoexcitation of a deaerated PhCN solution of ZnP–NIm resulted in ET from $^1ZnP^*$ to NIm to produce the CS state ($ZnP^{\bullet+}$–$NIm^{\bullet-}$) with the rate constant of $k_{CS} = 3.4 \times 10^9\ s^{-1}$, followed by BET from $NIm^{\bullet-}$ to $ZnP^{\bullet+}$ (CR) with $k_{CR} = 7.7 \times 10^5\ s^{-1}$, which corresponds to the CS lifetime of 1.3 μs at 298 K.[105] In the presence of metal ions (M^{n+}), the CS process occurs mainly from the $^1ZnP^*$–NIm rather than the $^1ZnP^*$–NIm/M^{n+} complex due to very weak binding of M^{n+} with neutral species of NIm.[105] In such a case, the k_{CS} values in the presence of M^{n+} are determined by the $-\Delta G_{CS}$ value in the absence of M^{n+} rather than the $-\Delta G_{CS}$ values in the presence of M^{n+}, although the $-\Delta G_{CS}$ values increase with increasing concentration of M^{n+}.[105] In contrast to this, the rate constants of BET (k'_{CR}: charge recombination) in the $ZnP^{\bullet+}$–$Nim^{\bullet-}$/M^{n+} complex are determined by the ΔG_{CR} values with strong binding of M^{n+}, since the CR process occurs mainly in the $ZnP^{\bullet+}$–$NIm^{\bullet-}$/M^{n+} complex rather than $ZnP^{\bullet+}$–$NIm^{\bullet-}$.[105] The BET from $NIm^{\bullet-}$ to $ZnP^{\bullet+}$ (CR) in the presence of Sc^{3+} (1.0×10^{-3} M) occurred with the k'_{CR} value of $6.9 \times 10^4\ s^{-1}$,

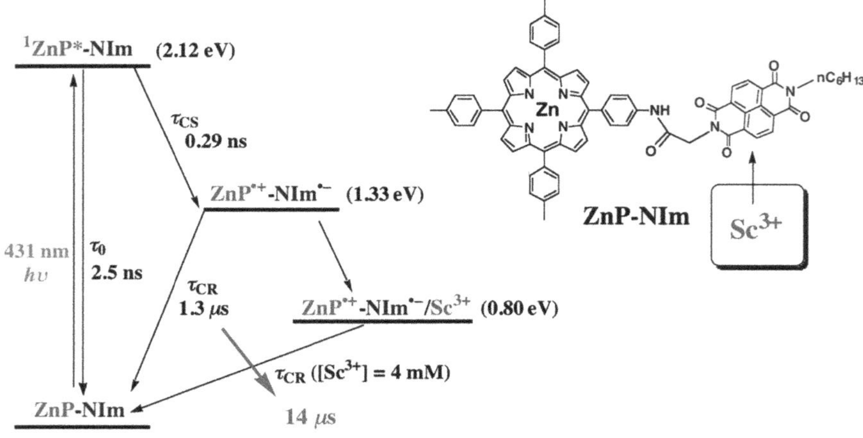

Scheme 3.11 Photoinduced ET and BET in ZnP–NIm in the absence and presence of a Sc^{3+} ion in PhCN. Reproduced from ref. 105 with permission from John Wiley and Sons Copyright © 2004 Wiley-VCH Verlag GmbH & Co. KGaA, Weinheim.

which corresponds to the CS lifetime of 14 μs.[105] The decrease in the driving force of BET (CR) together with an increase in the λ value due to the strong binding of M^{n+} to the $Nim^{•-}$ moiety of the CS state resulted in significant elongation of the CS lifetime.[105] The effect of the metal ion was the largest for Sc^{3+}, which is the strongest Lewis acid among metal ions, because the ΔG_{CR} value was the smallest for Sc^{3+}.[106,107]

Remarkable elongation of the CS lifetime has been achieved by binding of yttrium triflate [$Y(OTf)_3$] to the anthraquinone radical anion ($AQ^{•-}$) moiety of the CS state in photoinduced ET of a ferrocene-anthraquinone dyad (Fc–AQ) linked with a rigid amide spacer (Scheme 3.12).[108] Laser pulse excitation of the AQ moiety in Fc–AQ in deaerated PhCN resulted in ET from the Fc moiety to the $^1AQ^*$ moiety to produce the CS state as revealed by rise in the new transient absorption bands at 420 nm and 600 nm at 500 fs, which are assigned to $AQ^{•-}$ by comparison with the absorption spectrum of $AQ^{•-}$ produced by the ET reduction of AQ with naphthalene radical anion.[108] The decay of absorbance due to the CS state obeyed first-order kinetics with the lifetime of 12 ps.[108] Thus, ET from Fc to $^1AQ^*$ in Fc–AQ occurred rapidly to produce the CS state (Fc^+–$AQ^{•-}$) within 500 fs and decayed *via* BET to the ground state with a lifetime of 12 ps (Scheme 3.12a).[108]

Femtosecond laser pulse excitation of a deaerated PhCN solution of Fc–AQ in the presence of $Y(OTf)_3$ (1.0×10^{-2} M) also resulted in efficient ET from Fc to AQ within 500 fs.[108] However, the transient

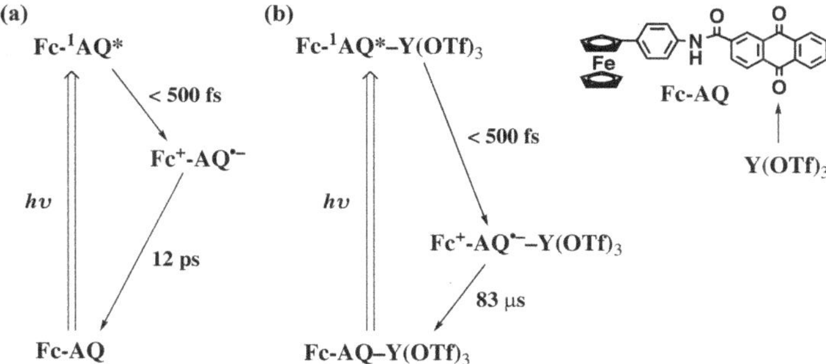

Scheme 3.12 Photoinduced ET and BET in a ferrocene–anthraquinone dyad linked with a rigid amide spacer (Fc–AQ) (a) in the absence and (b) in the presence of Y(OTf)$_3$. Reproduced from ref. 108 with permission from American Chemical Society, Copyright 2004.

absorption band observed at 700 nm in the presence of Y(OTf)$_3$ is significantly red-shifted as compared with that observed at 600 nm in the absence of Y(OTf)$_3$.[108] The same absorption band at 700 nm is observed in intermolecular photoinduced ET from Fc to AQ-ref (anthraquinone-2-carboxylic acid) in the presence of Y(OTf)$_3$.[108] Thus, the absorption band at 700 nm in the presence of Y(OTf)$_3$ is assigned due to the Fc$^+$–AQ$^{•-}$–Y(OTf)$_3$ complex in which Y(OTf)$_3$ binds with the oxygen atom of the AQ$^{•-}$ moiety strongly.[108] The decay of absorbance at 420 and 700 nm due to the Fc$^+$–AQ$^{•-}$–Y(OTf)$_3$ complex obeyed first-order kinetics to afford the CS lifetime of 83 μs (Scheme 3.12b), which is seven million times longer than the lifetime in the absence of Y(OTf)$_3$ (12 ps).[108] Such remarkable elongation of the CS lifetime may result from the strong binding of Y(OTf)$_3$ with AQ$^{•-}$, which causes a drastic increase in the reorganization energy as well as a significant decrease in the driving force of BET.[108]

The stabilization of the CS state by binding of metal ions can change the reaction pathway of an electron donor–acceptor dyad from an energy transfer pathway to an ET pathway (*vide infra*). Laser excitation at 530 nm of zinc tri-*tert*-butylaminoethyloxyphthalocyanine linked with *N*-(1-hexylheptyl)perylene-3,4,9,10-tetracarboxylic-3,4-anhydride-9,10-imide (ZnPc–PDI), where PDI has the absorption band, resulted in formation of the triplet excited state of ZnPc (^{3}ZnPc*–PDI) in deaerated benzonitrile (PhCN).[109] The triplet–triplet (T–T) absorption at 510 nm due to ^{3}ZnPc*–PDI was observed first. Then, it was changed to the T–T absorption at 700 nm due to ZnPc–^{3}PDI*, which agrees with the T–T absorption of ^{3}PDI* (Scheme 3.13a).[109] The rate constant of energy transfer from ^{3}ZnPc* to the PDI moiety was determined to be

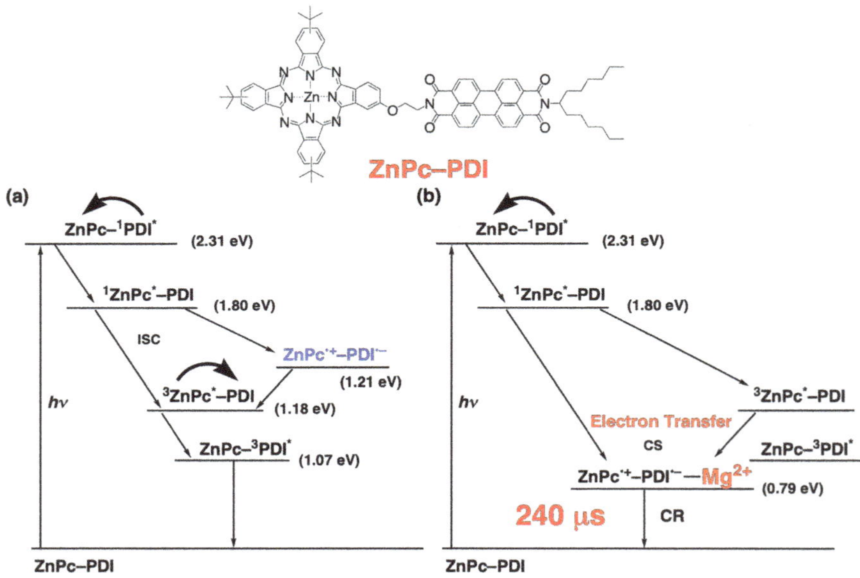

Scheme 3.13 Photoinduced EN, ET and BET in ZnPc–PDI (a) in the absence and (b) the presence of Mg^{2+} in PhCN. Reproduced from ref. 109 with permission from the Royal Society of Chemistry.

2.0×10^4 s^{-1} in deaerated PhCN at 298 K.[109] In the presence of $Mg(ClO_4)_2$ (0.10 M), ET from ^{3}ZnP* to PDI occurred instead of energy transfer from ^{3}ZnP* to PDI to produce the CS state, $ZnPc^{\bullet+}$–$PDI^{\bullet-}$/Mg^{2+}, which exhibits absorption bands at 550 nm due to $ZnPc^{\bullet+}$ together at 500 nm due to the $PDI^{\bullet-}/Mg^{2+}$ complex.[109] The decay of $ZnPc^{\bullet+}$–$PDI^{\bullet-}/Mg^{2+}$ obeyed first-order kinetics with the same slope due to intramolecular BET from the $PDI^{\bullet-}/Mg^{2+}$ complex to the $ZnPc^{\bullet+}$ moiety to afford a CS lifetime of 240 μs in Ar-saturated PhCN at 298 K.[109] The change in the reaction pathway from energy transfer (EN) to ET is caused by binding of Mg^{2+} to the $PDI^{\bullet-}$ moiety of $ZnPc^{\bullet+}$–$PDI^{\bullet-}$ to stabilize the CS state as compared with ^{3}PDI*.[109]

Similarly, laser photoexcitation of a bis(zinc phthalocyanine)–perylenediimide [$(ZnPc)_2$–PDI] triad resulted in the formation of ^{3}PDI*, whereas addition of Mg^{2+} ions to the triad resulted in formation of a long-lived CS state [$(ZnPc)_2^{\bullet+}$–$PDI^{\bullet-}/Mg^{2+}$) with a CS lifetime of 270 μs in Ar-saturated PhCN at 298 K.[110] Elongation of the CS lifetime by binding of metal ions has also been reported for a quinone-fused, bis-zinc porphyrin, ZnP–Q–ZnP,[111] zinc phthalocyanine-perylenebisimide pentameric arrays, $ZnPc(PDI)_4$,[112] and a zinc–quinoxalinoporphyrin/gold–quinoxalinoporphyrin dyad, ZnPA–AuPQ.[113]

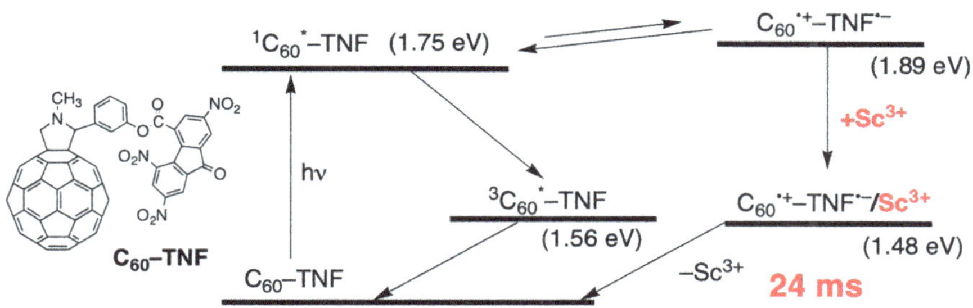

Scheme 3.14 Photoinduced ET and BET in a fullerene-trinitrofluorenone dyad (C_{60}–TNF) in the absence and presence of Sc^{3+}. Reproduced from ref. 116 with permission from the Royal Society of Chemistry.

The longest CS lifetime (110 ms) was obtained for a Pd porphyrin–flavin-linked dyad (PdP–Fl) to which two Sc^{3+} ions are bound to the flavin moiety upon photoexcitation of PdP–Fl/$(Sc^{3+})_2$ in deaerated PhCN at 298 K.[114] Binding of Sc^{3+} ions to Fl is reported to shift the one-electron reduction potential to a largely positive direction.[115]

Fullerene (C_{60}) has normally been used as an electron acceptor in electron donor–acceptor linked molecules. Binding of Sc^{3+} to an electron acceptor moiety (trinitrofluorenone) in a fullerene–trinitrofluorenone dyad (C_{60}–TNF) has enabled the use of the C_{60} moiety as an electron donor (Scheme 3.14).[116] No photoinduced ET occurred from the C_{60} moiety to the TNF moiety in C_{60}–TNF, because the energy of $^1C_{60}^*$ (1.75 eV) is lower than the CS state energy (1.89 eV).[116] In the presence of Sc^{3+} (30 mM), however, the CS state energy (1.48 eV) became lower than the energy of $^1C_{60}^*$ (1.75 eV) because of binding of Sc^{3+} to the $TNF^{•-}$ moiety (Scheme 3.14). In such a case, the CS state, $C_{60}^{•+}$–$TNF^{•-}$/Sc^{3+}, was produced upon photoexcitation of C_{60}–TNF in the presence of Sc^{3+} (30 mM) in PhCN.[116] The appearance of the absorption band at 960 nm in the laser-induced transient absorption spectra obtained upon photoexcitation of a deaerated PhCN solution of C_{60}–TNF in the presence of Sc^{3+} is a clear indication of formation of $C_{60}^{•+}$.[117] The $C_{60}^{•+}$ moiety acts as a strong electron acceptor (E_{red} *vs.* SCE = 1.49 V) to be capable of oxidizing *trans*-stilbene (E_{ox} *vs.* SCE = 1.47 V) producing a *trans*-stilbene radical cation.[117]

References

1. J. Deisenhofer and H. Michel, *Science*, 1989, **245**, 1463–1473.
2. A. J. Hoff and J. Deisenhofer, *Phys. Rep.*, 1997, **287**, 1–247.
3. J. J. Warren, J. R. Winkler and H. B. Gray, *Coord. Chem. Rev.*, 2013, **257**, 165–170.

4. L. C. Williams and A. K. W. Taguchi, in *Anoxygenic Photosynthetic Bacteria*, ed. R. E. Blankenship, M. T. Madigan and C. E. Bauer, Kluwer, Dordrecht, 1995, p. 1029.
5. *The Photosynthetic Reaction Center*, ed. J. Deisenhofer and J. R. Norris, Academic Press, San Diego, 1993.
6. U. Ermler, G. Fritzsch, S. K. Buchanan and H. Michel, *Structure*, 1994, **2**, 925–936.
7. S. G. Boxer, *Annu. Rev. Biophys. Biophys. Chem.*, 1990, **19**, 267–299.
8. K. M. Giangiacomo and P. L. Dutton, *Proc. Natl. Acad. Sci. U. S. A.*, 1989, **86**, 2658–2662.
9. S. Fukuzumi, in *The Porphyrin Handbook*, ed. K. M. Kadish, K. Smith and R. Guilard, Academic Press, San Diego, 2000, vol. 8, pp. 115–151.
10. S. Fukuzumi and H. Imahori, in *Electron Transfer in Chemistry*, ed. V. Balzani, Wiley-VCH, Weinheim, 2001, vol. 2, pp. 927–975.
11. S. Fukuzumi and D. M. Guldi, in *Electron Transfer in Chemistry*, ed. V. Balzani, Wiley-VCH, 2001, vol. 2, pp. 270–337.
12. D. M. Guldi and S. Fukuzumi, in *Fullerenes: Fullerenes:From Synthesis to Opto-electronic Properties*, ed. D. M. Guldi and N. Martin, Kluwer, Dordrecht, 2003, pp. 237–265.
13. S. Fukuzumi, K. Ohkubo, H. Imahori, J. Shao, Z. Ou, G. Zheng, Y. Chen, R. K. Pandey, M. Fujitsuka, O. Ito and K. M. Kadish, *J. Am. Chem. Soc.*, 2001, **123**, 10676–10683.
14. N. Holmberg-Douglas and D. A. Nicewicz, *Chem. Rev.*, 2022, **122**, 1925–2016.
15. H. Imahori, K. Tamaki, D. M. Guldi, C. Luo, M. Fujitsuka, O. Ito, Y. Sakata and S. Fukuzumi, *J. Am. Chem. Soc.*, 2001, **123**, 2607–2617.
16. K. Ohkubo, H. Kotani, J. Shao, Z. Ou, K. M. Kadish, G. Li, R. K. Pandey, M. Fujitsuka, O. Ito, H. Imahori and S. Fukuzumi, *Angew. Chem., Int. Ed.*, 2004, **43**, 853–856.
17. Y. Kashiwagi, K. Ohkubo, J. A. McDonald, I. M. Blake, M. J. Crossley, Y. Araki, O. Ito, H. Imahori and S. Fukuzumi, *Org. Lett.*, 2003, **5**, 2719–2721.
18. D. M. Guldi and P. V. Kamat, in *Fullerenes, Chemistry, Physics, and Technology*, ed. K. M. Kadish and R. S. Ruoff, Wiley-Interscience, New York, 2000, pp. 225–281.
19. S. Fukuzumi, K. Ohkubo, W. E. Z. Ou, J. Shao, K. M. Kadish, J. A. Hutchison, K. P. Ghiggino, P. J. Sintic and M. J. Crossley, *J. Am. Chem. Soc.*, 2003, **125**, 14984–14985.
20. M. Ohtani, K. Saito and S. Fukuzumi, *Chem. – Eur. J.*, 2009, **15**, 9160–9168.
21. K. Saito, M. Ohtani, T. Sakata, H. Mori and S. Fukuzumi, *J. Am. Chem. Soc.*, 2006, **128**, 14216–14217.
22. H. Imahori, D. M. Guldi, K. Tamaki, Y. Yoshida, C. Luo, Y. Sakata and S. Fukuzumi, *J. Am. Chem. Soc.*, 2001, **123**, 6617–6628.
23. H. Imahori, Y. Sekiguchi, Y. Kashiwagi, T. Sato, Y. Araki, O. Ito, H. Yamada and S. Fukuzumi, *Chem. – Eur. J.*, 2004, **10**, 3184–3196.
24. D. M. Guldi, H. Imahori, K. Tamaki, Y. Kashiwagi, H. Yamada, Y. Sakata and S. Fukuzumi, *J. Phys. Chem. A*, 2004, **108**, 541–548.
25. D. Gust, T. A. Moore and A. L. Moore, *Acc. Chem. Res.*, 1993, **26**, 198–205.
26. M. R. Wasielewski, *Chem. Rev.*, 1992, **92**, 435–461.
27. A. Osuka, N. Mataga and T. Okada, *Pure Appl. Chem.*, 1997, **69**, 797–802.
28. S. Fukuzumi, *Org. Biomol. Chem.*, 2003, **1**, 609–620.
29. D. Gust, T. A. Moore and A. L. Moore, *Acc. Chem. Res.*, 2001, **34**, 40–48.
30. S. Fukuzumi, *Phys. Chem. Chem. Phys.*, 2008, **10**, 2283–2297.
31. S. Fukuzumi, K. Ohkubo and T. Suenobu, *Acc. Chem. Res.*, 2014, **47**, 1455–1464.
32. M. Rudolf, S. V. Kirner and D. M. Guldi, *Chem. Soc. Rev.*, 2016, **45**, 612–630.
33. H. Imahori, *Bull. Chem. Soc. Jpn.*, 2023, **96**, 339–352.
34. C. Wang, B. Wu and C. Wang, *Acc. Mater. Res.*, 2024, **5**, 426–437.
35. A. Helms, D. Heiler and G. McLendon, *J. Am. Chem. Soc.*, 1992, **114**, 6227–6238.

36. S. Fukuzumi, K. Ohkubo, T. Suenobu, K. Kato, M. Fujitsuka and O. Ito, *J. Am. Chem. Soc.*, 2001, **123**, 8459–8467.
37. S. Fukuzumi, H. Kotani, K. Ohkubo, S. Ogo, N. V. Tkachenko and H. Lemmetyinen, *J. Am. Chem. Soc.*, 2004, **126**, 1600–1601.
38. K. Ohkubo, H. Kotani and S. Fukuzumi, *Chem. Commun.*, 2005, 4520–4522.
39. M. Hoshino, H. Uekusa, A. Tomita, S. Koshihara, T. Sato, S. Nozawa, S. Adachi, K. Ohkubo, H. Kotani and S. Fukuzumi, *J. Am. Chem. Soc.*, 2012, **134**, 4569–4572.
40. S. Fukuzumi, K. Doi, A. Itoh, T. Suenobu, K. Ohkubo, Y. Yamada and K. D. Karlin, *Proc. Natl. Acad. Sci. U. S. A.*, 2012, **109**, 15572–15577.
41. T. Tsudaka, H. Kotani, K. Ohkubo, T. Nakagawa, N. V. Tkachenko, H. Lemmetyinen and S. Fukuzumi, *Chem. – Eur. J.*, 2017, **23**, 1306–1317.
42. K. Ohkubo, K. Suga, K. Morikawa and S. Fukuzumi, *J. Am. Chem. Soc.*, 2003, **125**, 12850–12859.
43. T. Shida and S. Iwata, *J. Am. Chem. Soc.*, 1973, **95**, 3473–3483.
44. K. Ohkubo, S. Matsumoto, H. Asahara and S. Fukuzumi, *ACS Catal.*, 2024, **14**, 2671–2684.
45. T. Umemoto and S. Ishimara, *Tetrahedron Lett.*, 1990, **31**, 3579–3582.
46. T. Umemoto, *Chem. Rev.*, 1996, **96**, 1757–1778.
47. S. Fukuzumi, Y. Yamada, T. Suenobu, K. Ohkubo and H. Kotani, *Energy Environ. Sci.*, 2011, **4**, 2754–2766.
48. S. Fukuzumi and K. Ohkubo, *Chem. Sci.*, 2013, **4**, 561–574.
49. S. Fukuzumi and K. Ohkubo, *Org. Biomol. Chem.*, 2014, **12**, 6059–6071.
50. K. A. Margrey and D. A. Nicewicz, *Acc. Chem. Res.*, 2016, **49**, 1997–2006.
51. N. A. Romero and D. A. Nicewicz, *Chem. Rev.*, 2016, **116**, 10075–10166.
52. P. K. Verma, *Coord. Chem. Rev.*, 2022, **472**, 214805.
53. C.-Y. Huang, J. Li and C.-J. Li, *Chem. Sci.*, 2022, **13**, 5465–5504.
54. D. Nicewicz, *Synlett*, 2022, **33**, 1135–1136.
55. Y.-M. Lee, W. Nam and S. Fukuzumi, *Chem. Sci.*, 2023, **14**, 4205–4218.
56. L. Qian and M. Shi, *Chem. Commun.*, 2023, **59**, 3487–3506.
57. L. Chang, S. Wang, Q. An, L. Liu, H. Wang, Y. Li, K. Feng and Z. Zuo, *Chem. Sci.*, 2023, **14**, 6841–6859.
58. J. Zhang and M. Rueping, *Chem. Soc. Rev.*, 2023, **52**, 4099–4120.
59. H. Kotani, K. Ohkubo and S. Fukuzumi, *Faraday Discuss.*, 2012, **155**, 89–102.
60. K. Ohkubo, K. Mizushima, R. Iwata, K. Souma, N. Suzuki and S. Fukuzumi, *Chem. Commun.*, 2010, **46**, 601–603.
61. Y. Yamada, A. Nomura, K. Ohkubo, T. Suenobu and S. Fukuzumi, *Chem. Commun.*, 2013, **49**, 5132–5134.
62. Y. Yamada, T. Miyahigashi, H. Kotani, K. Ohkubo and S. Fukuzumi, *J. Am. Chem. Soc.*, 2011, **133**, 16136–16145.
63. S. Fukuzumi, S. Koumitsu, K. Hironaka and T. Tanaka, *J. Am. Chem. Soc.*, 1987, **109**, 305–316.
64. Y. Yamada, T. Miyahigashi, H. Kotani, K. Ohkubo and S. Fukuzumi, *Energy Environ. Sci.*, 2012, **5**, 6111–6118.
65. F. D'Souza and O. Ito, *Coord. Chem. Rev.*, 2005, **249**, 1410–1422.
66. S. Fukuzumi and T. Kojima, *J. Mater. Chem.*, 2008, **18**, 1427–1439.
67. F. D'Souza and O. Ito, *Chem. Commun.*, 2009, 4913–4928.
68. G. Bottari, O. Trukhina, M. Ince and T. Torres, *Coord. Chem. Rev.*, 2012, **256**, 2453–2477.
69. S. Fukuzumi, K. Ohkubo, F. D'Souza and J. L. Sessler, *Chem. Commun.*, 2012, **48**, 9801–9815.
70. N. L. Bill, O. Trukhina, J. L. Sessler and T. Torres, *Chem. Commun.*, 2015, **51**, 7781–7794.
71. J. F. Nierengarten, *Eur. J. Inorg. Chem.*, 2019, 4865–4878.
72. S. Das and M. Presselt, *J. Mater. Chem. C*, 2019, **7**, 6194–6216.

73. X. Chang, Y. Xu and M. Von Delius, *Chem. Soc. Rev.*, 2024, **53**, 47–83.
74. M. Tanaka, K. Ohkubo, C. P. Gros, R. Guilard and S. Fukuzumi, *J. Am. Chem. Soc.*, 2006, **128**, 14625–14633.
75. E. J. Dale, N. A. Vermeulen, M. Juríc?ek, J. C. Barnes, R. M. Young, M. R. Wasielewski and J. F. Stoddart, *Acc. Chem. Res.*, 2016, **49**, 262–273.
76. S. Fukuzumi, I. Amasaki, K. Ohkubo, C. P. Gros, R. Guilard and J.-M. Barbe, *RSC Adv.*, 2012, **2**, 3741–3747.
77. F. Diederich and M. Gómez-López, *Chem. Soc. Rev.*, 1999, **28**, 263–277.
78. D. Sun, F. S. Tham, C. A. Reed, L. Chaker and P. D. W. Boyd, *J. Am. Chem. Soc.*, 2002, **124**, 6604–6612.
79. S. K. Samanta and M. Schmittel, *Org. Biomol. Chem.*, 2013, **11**, 3108–3115.
80. K. Börjesson, J. G. Woller, E. Parsa, J. Mårtensson and B. Albinsson, *Chem. Commun.*, 2012, **48**, 1793–1795.
81. B. Kang, R. K. Totten, M. H. Weston, J. T. Hupp and S. T. Nguyen, *Dalton Trans.*, 2012, **41**, 12156–12162.
82. G. Wu, F. Li, B. Tang and X. Zhang, *J. Am. Chem. Soc.*, 2022, **144**, 14962–14975.
83. H. Nobukuni, F. Tani, Y. Shimazaki, Y. Naruta, K. Ohkubo, T. Nakanishi, T. Kojima, S. Fukuzumi and S. Seki, *J. Phys. Chem. C*, 2009, **113**, 19694–19699.
84. H. Nobukuni, Y. Shimazaki, H. Uno, Y. Naruta, K. Ohkubo, T. Kojima, S. Fukuzumi, S. Seki, H. Sakai, T. Hasobe and F. Tani, *Chem. – Eur. J.*, 2010, **16**, 11611–11623.
85. T. Kamimura, K. Ohkubo, Y. Kawashima, H. Nobukuni, Y. Naruta, F. Tani and S. Fukuzumi, *Chem. Sci.*, 2013, **4**, 1451–1461.
86. Y. Kawashima, K. Ohkubo and S. Fukuzumi, *J. Phys. Chem. A*, 2012, **116**, 8942–8948.
87. A. Takai, M. Chkounda, A. Eggenspiller, C. P. Gros, M. Lachkar, J.-M. Barbe and S. Fukuzumi, *J. Am. Chem. Soc.*, 2010, **132**, 4477–4489.
88. A. Takai, C. P. Gros, J.-M. Barbe and S. Fukuzumi, *Phys. Chem. Chem. Phys.*, 2010, **12**, 12160–12168.
89. N. Solladié, S. Fukuzumi, K. Ohkubo, F. D'Souza, R. Rein, K. Saito, V. Troiani, H. Qiu, S. Gadde and T. Hasegawa, *ECS J. Solid State Sci. Technol.*, 2020, **9**, 061026.
90. N. Solladié, A. Hamel and M. Gross, *Tetrahedron Lett.*, 2000, **41**, 6075–6078.
91. K. Saito, V. Troiani, H. Qiu, N. Solladié, T. Sakata, H. Mori, M. Ohama and S. Fukuzumi, *J. Phys. Chem. C*, 2007, **111**, 1194–1199.
92. S. Fukuzumi, K. Ohkubo, K. Saito, Y. Kashiwagi and M. J. Crossley, *J. Porphyrins Phthalocyanines*, 2011, **15**, 1292–1298.
93. T. Hasobe, Y. Kashiwagi, M. A. Absalom, J. Sly, K. Hosomizu, M. J. Crossley, H. Imahori, P. V. Kamat and S. Fukuzumi, *Adv. Mater.*, 2004, **16**, 975–979.
94. S. Fukuzumi, K. Saito, K. Ohkubo, T. Khoury, Y. Kashiwagi, M. A. Absalom, S. Gadde, F. D'Souza, Y. Araki, O. Ito and M. J. Crossley, *Chem. Commun.*, 2011, **47**, 7980–7982.
95. S. Aoyagi, E. Nishibori, H. Sawa, K. Sugimoto, M. Takata, Y. Miyata, R. Kitaura, H. Shinohara, H. Okada, T. Sakai, Y. Ono, K. Kawachi, K. Yokoo, S. Ono, K. Omote, Y. Kasama, S. Ishikawa, T. Komuro and H. Tobita, *Nat. Chem.*, 2010, **4**, 678–683.
96. K. Ohkubo, Y. Kawashima and S. Fukuzumi, *Chem. Commun.*, 2012, **48**, 4314–4316.
97. K. M. Kadish, R. K. Rhodes and L. A. Bottomleym, *Inorg. Chem.*, 1981, **20**, 1274–1277.
98. N. L. Bill, M. Ishida, Y. Kawashima, K. Ohkubo, Y. M. Sung, V. M. Lynch, J. M. Lim, D. Kim, J. L. Sessler and S. Fukuzumi, *Chem. Sci.*, 2014, **5**, 3888–3896.
99. J. L. Sessler, E. Karnas, S. K. Kim, Z. Ou, M. Zhang, K. M. Kadish, K. Ohkubo and S. Fukuzumi, *J. Am. Chem. Soc.*, 2008, **130**, 15256–15257.

100. S. Fukuzumi, N. Satoh, Y. Yuasa and K. Ohkubo, *J. Am. Chem. Soc.*, 2004, **126**, 7585–7594.
101. T. Honda, T. Nakanishi, K. Ohkubo, T. Kojima and S. Fukuzumi, *J. Am. Chem. Soc.*, 2010, **132**, 10155–10163.
102. R. A. Marcus and N. Sutin, *Biochim. Biophys. Acta, Rev. Bioenerg.*, 1985, **811**, 265–322.
103. F. D'Souza, E. Maligaspe, K. Ohkubo, M. E. Zandler, N. K. Subbaiyan and S. Fukuzumi, *J. Am. Chem. Soc.*, 2009, **131**, 8787–8797.
104. S. Fukuzumi and K. Ohkubo, *Coord. Chem. Rev.*, 2010, **254**, 372–385.
105. K. Okamoto, Y. Mori, H. Yamada, H. Imahori and S. Fukuzumi, *Chem. - Eur. J.*, 2004, **10**, 474–483.
106. S. Fukuzumi and K. Ohkubo, *Chem. - Eur. J.*, 2000, **6**, 4532–4535.
107. S. Fukuzumi and K. Ohkubo, *J. Am. Chem. Soc.*, 2002, **124**, 10270–10271.
108. K. Okamoto, Y. Araki, O. Ito and S. Fukuzumi, *J. Am. Chem. Soc.*, 2004, **126**, 56–57.
109. S. Fukuzumi, K. Ohkubo and Á. Sastre-Santos, *Chem. Commun.*, 2005, 3814–3816.
110. S. Fukuzumi, K. Ohkubo, J. Ortiz, A. M. Gutiérrez, F. Fernández-Lázaro and Á. Sastre-Santos, *J. Phys. Chem. A*, 2008, **112**, 10744–10752.
111. M. B. Thomas, Y. Hu, W. Shan, K. M. Kadish, H. Wang and F. D'Souza, *J. Phys. Chem. C*, 2019, **123**, 22066–22073.
112. K. Ohkubo, S. Fukuzumi, Á. Sastre-Santos and F. Fernández-Lázaro, *J. Org. Chem.*, 2009, **74**, 5871–5880.
113. K. Ohkubo, R. Garcia, P. J. Sintic, T. Khoury, M. J. Crossley, K. M. Kadish and S. Fukuzumi, *Chem. - Eur. J.*, 2009, **15**, 10493–10503.
114. T. Kojima, R. Kobayashi, T. Ishizuka, S. Yamakawa, H. Kotani, T. Nakanishi, K. Ohkubo, Y. Shiota, K. Yoshizawa and S. Fukuzumi, *Chem. - Eur. J.*, 2014, **20**, 15518–15532.
115. S. Fukuzumi, K. Yasui, T. Suenobu, K. Ohkubo, M. Fujitsuka and O. Ito, *J. Phys. Chem.*, 2001, **105**, 10501–10510.
116. K. Ohkubo, J. Ortiz, L. Martín-Gomis, F. Fernández-Lázaro, Á. Sastre-Santos and S. Fukuzumi, *Chem. Commun.*, 2007, 589–591.
117. S. Fukuzumi, H. Mori, H. Imahori, T. Suenobu, Y. Araki, O. Ito and K. M. Kadish, *J. Am. Chem. Soc.*, 2001, **123**, 12458–12465.

4 PSI Models

4.1 Hydrogen Production

Photosynthesis involves two photosystems, photosystem I (PSI) and photosystem II (PSII), in both of which the photosynthetic reaction centres undergo multi-step photoinduced electron transfer to attain long-lived charge separation.[1] In PSI, the positive charge oxidizes plastoquinol (PQH_2), whereas the negative charge reduces the $NADP^+$ coenzyme (*vide infra*).[2,3] When the negative charge reduces protons to evolve H_2, the stoichiometry of the PSI model, namely the photocatalytic evolution of H_2 by PQH_2 [eqn (4.1)], is achieved.[4,5] Extensive efforts have so far been made to develop efficient photocatalytic H_2 evolution using a variety of sacrificial electron donors.[6–18]

$$PQH_2 \xrightarrow{h\nu} H_2 \tag{4.1}$$

The 9-mesityl-10-methylacridinium ion (Acr^+–Mes) has been shown to act as an efficient photosynthetic centre (PRC) model compound, because photoexcitation of an Ar-saturated MeCN solution of Acr^+–Mes results in formation of a long-lived triplet ET state [$^3(Acr^\bullet–Mes^{\bullet+})$], which is capable of oxidizing electron donors and reducing electron acceptors at the same time (*vide supra*).[19,20] The first functional molecular model of PSI has been constructed, where hydroquinone derivatives (X-Q) were employed as plastoquinol analogues, Acr^+–Mes as an efficient PRC model and $Co^{III}(dmgH)_2pyCl$ as an H_2 evolution catalyst[21] to achieve the stoichiometry of PSI (Scheme 4.1).[22]

RSC Foundations No. 1
Artificial Photosynthesis
By Shunichi Fukuzumi

Published by the Royal Society of Chemistry, www.rsc.org

(a)

Plastoquinol

QH_2 Hydroquinone

Cl_4QH_2 Tetrachlorohydroquinone

Me_4QH_2 Durohydroquinone

Hydroquinone derivatives ($X\text{-}QH_2$)

(b)

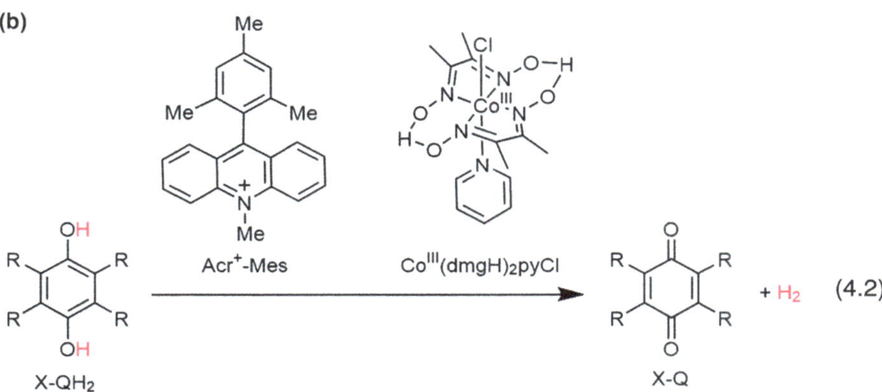

Scheme 4.1 (a) Chemical structures of plastoquinol (PQ) and $X\text{-}QH_2$ (PQ analogue). (b) Photocatalytic hydrogen evolution from $X\text{-}QH_2$ with Acr^+–Mes (a photosynthetic reaction centre (PRC) model) and $Co^{III}(dmgH)_2pyCl$ (H_2 evolution catalyst). Reproduced from ref. 22 with permission from American Chemical Society, Copyright 2020.

Visible light irradiation ($\lambda > 420$ nm) of an Ar-saturated MeCN solution containing a hydroquinone derivative ($X\text{-}QH_2$) (1.0×10^{-3} M), Acr^+–Mes (7.5×10^{-4} M), $Co^{III}(dmgH)_2pyCl$ (6.0×10^{-4} M) and H_2O (2.0 M) resulted in hydrogen (H_2) evolution, as shown in Figure 4.1, where the H_2 yield increased with irradiation time to reach 100% yield based on the initial amount of $X\text{-}QH_2$ [eqn (4.2) in Scheme 4.1b].[22] The absence of Acr^+–Mes or $Co^{III}(dmgH)_2pyCl$ or H_2O resulted in no H_2 evolution under visible light irradiation.[22] Thus, the presence of Acr^+–Mes and $Co^{III}(dmgH)_2pyCl$ are essential for the photocatalytic H_2 evolution from $X\text{-}QH_2$. The quantum yield for the photocatalytic H_2 evolution from Cl_4QH_2 (10 mM) with Acr^+–Mes (0.75 mM) and $Co^{III}(dmgH)_2pyCl$ (0.60 mM) in Ar-saturated MeCN containing H_2O (2.0 M) was determined to be 10% with use of a ferrioxalate actinometer.[22]

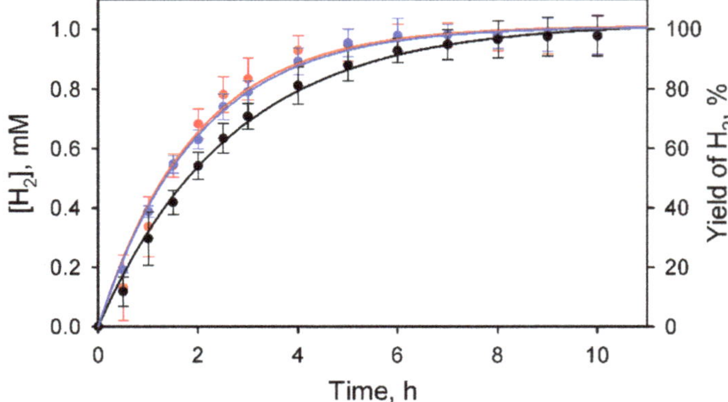

Figure 4.1 Time courses of H_2 evolution from X-QH_2 [(1.0×10^{-3} M): QH_2 (black circles), Cl_4QH_2 (red circles), and Me_4QH_2 (blue circles)] with Acr^+–Mes (7.5×10^{-4} M) and $Co^{III}(dmgH)_2pyCl$ (6.0×10^{-4} M) in deaerated MeCN containing H_2O (2.0 M) under visible light irradiation ($\lambda > 420$ nm) at 298 K. Reproduced from ref. 22 with permission from American Chemical Society, Copyright 2020.

When tetrachlorohydroquinone (Cl_4QH_2) was employed as a PQH_2 analogue, the absorption band at 287 nm due to formation of *p*-chloranil (Cl_4Q) increased to reach 100% yield in 6 h photo-irradiation.[22] The absorption spectrum of Acr^+–Mes ($\lambda_{max} = 360$ and 423 nm) remained virtually unchanged during the photocatalytic H_2 evolution from Cl_4QH_2 with Acr^+–Mes, $Co^{III}(dmgH)_2pyCl$, and H_2O for 6 h, suggesting that Acr^+–Mes is stable during the photocatalytic H_2 evolution.[22] In the case of $DDQH_2$ and Me_2QH_2, the hydroquinone derivatives were oxidized to the corresponding *p*-benzoquinone derivatives in the photocatalytic evolution of H_2.[22] Formation of *p*-benzoquinone (BQ) from hydroquinone (QH_2) in the photocatalytic H_2 evolution was also confirmed by ^{1}H NMR measurements in CD_3CN, in which NMR signals due to QH_2 decreased, accompanied by appearance of those due to BQ (Scheme 4.1).[22] Thus, photocatalytic evolution occurred by use of a series of hydroquinone derivatives (X-QH_2) to mimic the stoichiometry of the photoredox reaction in PSI to produce X-Q.[22] This is the first molecular model of PSI for H_2 evolution from PQH_2 analogues.

ET from Cl_4QH_2 to the $Mes^{\bullet+}$ moiety of $^3(AcrH^{\bullet}$–$Mes^{\bullet+})$ was monitored directly by laser-induced transient absorption measurements.[22] The decay rate of the absorption band at 500 nm due to the $Mes^{\bullet+}$ moiety of $^3(Acr^{\bullet}$–$Mes^{\bullet+})$, which obeyed first-order kinetics, increased with increasing concentration of Cl_4QH_2 (Figure 4.2a).[22]

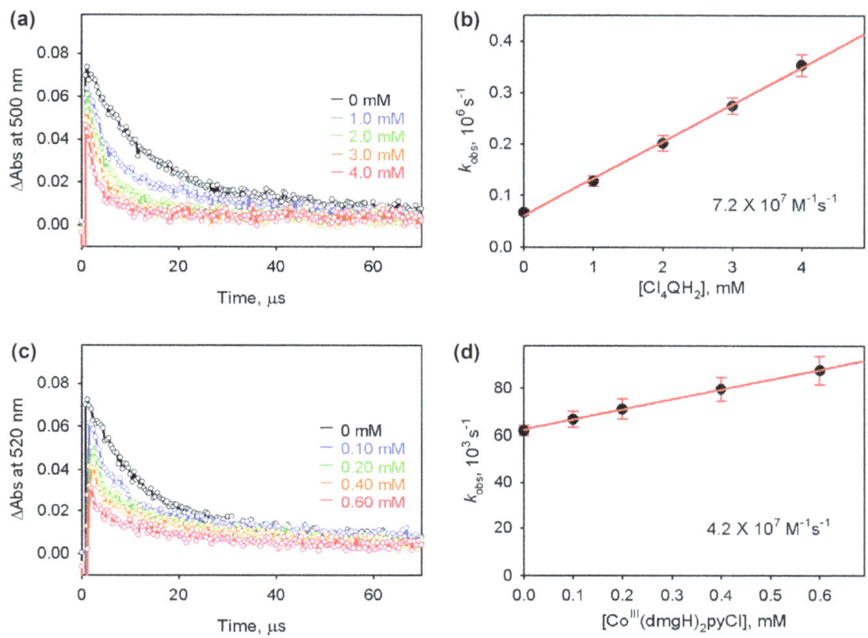

Figure 4.2 (a) Decay time courses of transient absorption at 500 nm due to the $Mes^{\bullet+}$ moiety of $^3(Acr^{\bullet}–Mes^{\bullet+})$ in the presence of various concentrations of Cl_4QH_2 [0 (black), 1.0 (blue), 2.0 (green), 3.0 (orange), and 4.0 (red) mM] after laser photoexcitation ($\lambda_{ex} = 355$ nm) of a deaerated MeCN solution of Acr^+–Mes (1.0×10^{-4} M). (b) Plot of the observed first-order rate constant (k_{obs}) *vs.* concentration of Cl_4QH_2. (c) Decay time courses of transient absorption at 520 nm due to the decay of $Acr^{\bullet}$ moiety of $Acr^{\bullet}–Mes^{\bullet+}$ in the presence of various concentrations of $Co^{III}(dmgH)_2pyCl$ [0 (black), 0.10 (blue), 0.10 (green), 0.30 (orange), and 0.60 (red) mM] after laser photoexcitation at 355 nm of a deaerated MeCN solution of Acr^+–Mes (0.10 mM) and Cl_4QH_2 (1.0×10^{-3} M). (d) Plot of the observed first-order rate constant (k_{obs}) *vs.* concentration of $Co^{III}(dmgH)_2pyCl$. Reproduced from ref. 22 with permission from American Chemical Society, Copyright 2020.

The second-order rate constant of ET from Cl_4QH_2 to the $Mes^{\bullet+}$ moiety of $^3(Acr^{\bullet}–Mes^{\bullet+})$ was determined from the slope of the first-order decay rate constant *vs.* concentration of Cl_4QH_2 to be 7.2×10^7 M^{-1} s^{-1} in deaerated MeCN at 298 K (Figure 4.2b).[22] ET from Cl_4QH_2 (E_{pa} *vs.* SCE = 1.05 V) to the $Mes^{\bullet+}$ moiety of $^3(Acr^{\bullet}–Mes^{\bullet+})$ (E_{red} *vs.* SCE = 2.06 V)[23] is highly exergonic.[22] The second-order rate constant of ET from the $Acr^{\bullet}$ moiety of $^3(Acr^{\bullet}–Mes^{\bullet+})$ to $Co^{III}(dmgH)_2pyCl$ was also determined from the slope of the first-order decay rate constant (Figure 4.2c) *vs.* concentration of Cl_4QH_2 to be 4.2×10^7 M^{-1} s^{-1} in deaerated MeCN at 298 K (Figure 4.2d).[22]

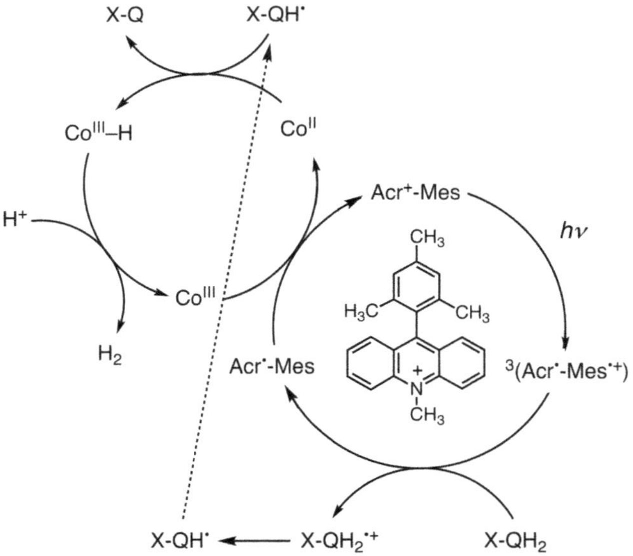

Scheme 4.2 Mechanism of photocatalytic H_2 evolution from QH_2 with Acr^+–Mes (a PRC model) and $Co^{III}(dmgH)_2pyCl$ (H_2 evolution catalyst).

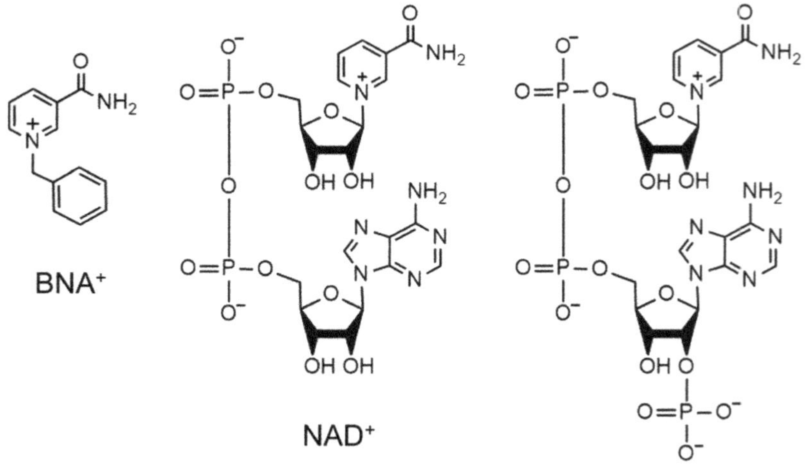

Figure 4.3 Chemical structures of NAD^+ analogues (BNA^+, NAD^+ and $NADP^+$).

ET from the Acr• moiety of ^{3}(Acr•–Mes•+) (E_{ox} *vs.* SCE = −0.57 V) to $Co^{III}(dmgH)_2pyCl$ (E_{red} *vs.* SCE = −0.16 V) is also exergonic.[22] Thus, Acr^+–Mes acts as a PRC (photosynthetic reaction centre) in a PSI model to show the capability to oxidize and reduce substrates at the same time.[22] Hydroquinone radical cation ($QH_2^{\bullet +}$) produced by ET from QH_2 to the $Mes^{\bullet +}$ moiety of $^3(Acr^{\bullet}–Mes^{\bullet +})$ deprotonates rapidly to produce a semiquinone radical ($QH^{\bullet}$), which was detected by transient absorption measurements. Semiquinone radicals $QH^{\bullet}$, $Cl_4QH^{\bullet}$ and $Me_4QH^{\bullet}$ were detected at $\lambda_{max} = 400$ nm, 410 nm and 430 nm, respectively.[22]

The mechanism of the photocatalytic H_2 evolution from X-QH_2 with Acr^+–Mes and $Co^{III}(dmgH)_2pyCl$ (PSI model) is summarized in Scheme 4.2.[22] As discussed above, photoexcitation of Acr^+–Mes (a PRC model) results in formation of the ET state of Acr^+–Mes [$^3(Acr^{\bullet}–Mes^{\bullet +})$], which is capable of oxidizing X-QH_2 to X-$QH_2^{\bullet +}$ and reducing $Co^{III}(dmgH)_2pyCl$ to $[Co^{II}(dmgH)_2pyCl]^-$, accompanied by regeneration of Acr^+–Mes.[22] X-$QH_2^{\bullet +}$ is deprotonated rapidly to produce

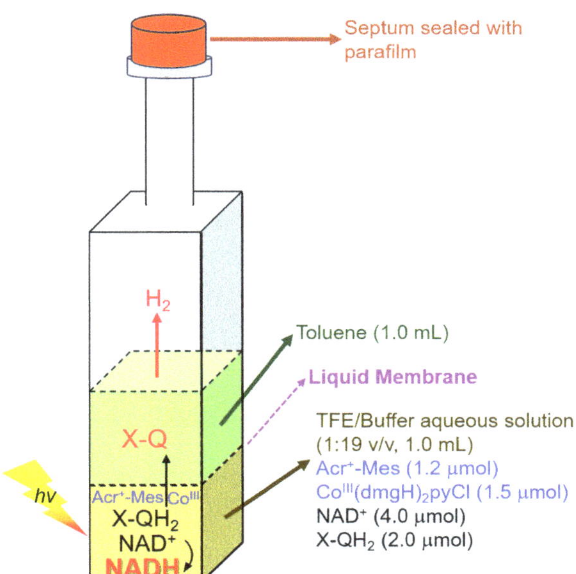

Figure 4.4 A quartz cell employed for the PSI model reaction: Photocatalytic reduction of NAD^+ by X-QH_2 in the toluene phase (1.0 mL; upper part) and the solvent mixture phase (1.0 mL; lower part) of borate buffer aqueous solution (0.10 M, 0.95 mL, pH 7.0) and TFE (0.050 mL) (19 : 1, v/v) containing X-QH_2 (2.0 μmol), NAD^+ (4.0 μmol), Acr^+–Mes (1.5 μmol, 0.020 mL of TFE) and $Co^{III}(dmgH)_2pyCl$ (1.2 μmol, 0.030 mL of TFE) to produce NADH regioselectively (only 1,4-dihydroform was produced). The irradiation was focused on the aqueous/TFE phase in which Acr^+–Mes (a PRC) is dissolved. Reproduced from ref. 25 with permission from American Chemical Society, Copyright 2024.

semiquinone radical (X-QH$^•$).[24] Then, hydrogen atom transfer or proton-coupled ET (PCET) from X-QH$^•$ to $[Co^{II}(dmgH)_2pyCl]^-$ occurs to produce X-Q and the Co(III)-hydride complex $[Co^{III}(H)(dmgH)_2PyCl]^-$, which reacts with H^+ to evolve H_2, accompanied by regeneration of $Co^{III}(dmgH)_2pyCl$.[22] When H_2O was replaced by D_2O, the H_2 evolution rate became slower to afford the kinetic deuterium isotope (KIE = 2.0).[22] The KIE and the observation of $[Co^{III}(H)(dmgH)_2pyCl]^-$ (g = [2.2500, 2.2500, 1.9950] with hyperfine coupling tensor A_{Co} = [28, 28, 302] MHz) suggests that hydrogen atom transfer or proton-coupled ET (PCET) from X-QH$^•$ to $[Co^{II}(dmgH)_2pyCl]^-$ is the rate-determining step in the photocatalytic H_2 revolution in Scheme 4.2.[22]

4.2 Reduction of NAD(P)$^+$ to NAD(P)H by PQ Analogues

In photosynthesis, PQH_2 reduced NAD(P)$^+$ regioselectively to produce NAD(P)H (1,4-dihydro form) *via* charge separation in the PRC in PSI [eqn (4.3)].[1] NAD$^+$ and its analogues (BNA$^+$ and NADP$^+$) are shown in Figure 4.3. The first molecular model of PSI for NAD(P)H production was reported by use of X-QH_2 as an electron and proton source to reduce NAD$^+$ regioselectively to produce NADH (1,4-dihydro form) in the presence of 9-mesityl-10-methylacridinium ion (Acr$^+$–Mes) as a PRC model and $Co^{III}(dmgH)_2pyCl$ as a NAD$^+$ reduction catalyst under visible light irradiation [eqn (4.4)].[25]

$$\mathrm{PQH_2 + NAD(P)^+ \xrightarrow[\text{PSI}]{h\nu} PQ + NAD(P)H + H^+} \tag{4.3}$$

$$\mathrm{X\text{-}QH_2 + NAD^+ \xrightarrow[\text{Acr}^+\text{-Mes,Co}^{III}(\text{dmgH})_2\text{pyCl}]{h\nu} X\text{-}Q + NADH + H^+} \tag{4.4}$$

A quartz cell was used for the PSI model reaction [eqn (4.4)] in the two phases separated by a liquid membrane under visible light irradiation (Figure 4.4).[25] Formation of NADH by the photocatalytic reduction of NAD$^+$ by Cl_4QH_2 was observed by HPLC measurements, whereas evolution of H_2 was detected by GC (Figure 4.5a).[25] NAD$^+$ is reduced regioselectively to NADH, which is converted to H_2 at the later state of the reaction (Figure 4.5a), eqn (4.5).[25] The conversion yield of NADH to H_2 was nearly 20%.[25]

$$\mathrm{NADH + H^+ \xrightarrow[\text{Acr}^+\text{-Mes,Co}^{III}(\text{dmgH})_2\text{pyCl}]{h\nu} NAD^+ + H_2} \tag{4.5}$$

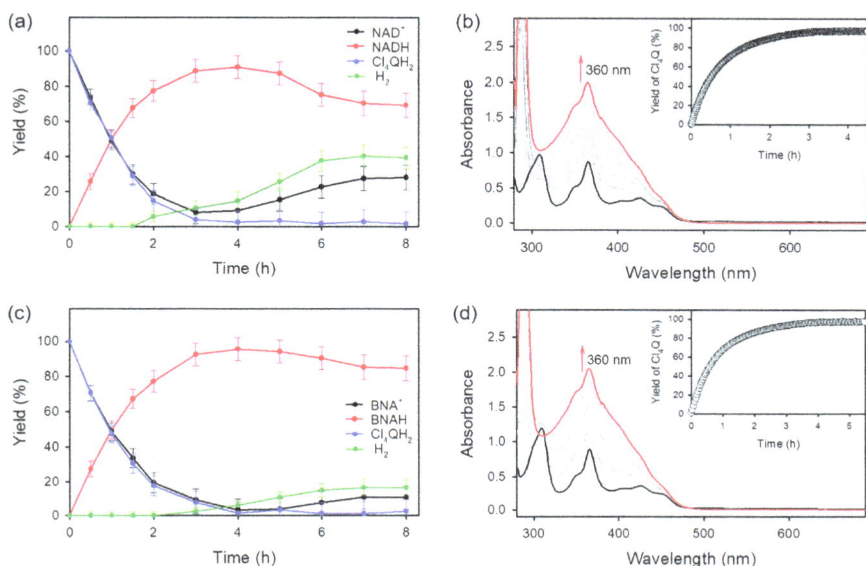

Figure 4.5 (a) and (c) Time courses of a PSI model reaction: photocatalytic reduction of NAD^+ analogues [(a) NAD^+ and (c) BNA^+; 4.0 μmol] or H^+ to produce NADH analogues [(a) NADH and (c) BNAH] or H_2 by Cl_4QH_2 (2.0 μmol) with a PRC model compound (Acr^+–Mes: 1.5 μmol) and $Co^{III}(dmgH)_2pyCl$ (NAD^+ reduction catalyst: 1.2 μmol) in the toluene phase (1.0 mL) combined by a liquid membrane with the solvent mixture phase (1.0 mL; lower part) of borate buffer aqueous solution (0.10 M, 0.95 mL, pH 7.0) and TFE (0.050 mL) (19 : 1, v/v) containing X-QH_2 (2.0 μmol), NAD^+ (4.0 μmol), Acr^+–Mes (1.5 μmol, 0.020 mL of TFE) and $Co^{III}(dmgH)_2pyCl$ (1.2 μmol, 0.030 mL of TFE) to produce NADH regioselectively. (b) and (d) UV-vis absorption spectral changes of the toluene phase observed in the photocatalytic reduction of NAD^+ analogues [(b) NAD^+ and (d) BNA^+; 4.0 μmol] by Cl_4QH_2 (2.0 μmol) with Acr^+–Mes (1.5 μmol) and $Co^{III}(dmgH)_2pyCl$ (1.2 μmol) in a toluene/TFE/borate buffer aqueous solution (0.10 M, pH 7.0) (3.0 mL, 40 : 1 : 19 v/v/v) at 298 K. Inset shows time profile of Cl_4Q yield. Reproduced from ref. 25 with permission from American Chemical Society, Copyright 2024.

The quantitative formation of Cl_4Q [eqn (4.4)] was also observed by UV-vis absorption spectral changes as shown in Figure 4.5b.[25] When NAD^+ was replaced by an NAD^+ model compound (BNA^+), BNAH was also selectively produced (Figure 4.5c), accompanied by formation of Cl_4Q (Figure 4.5d).[25]

The mechanism of a PSI model reaction (photocatalytic reduction of NAD^+ by X-QH_2 with Acr^+–Mes (a PRC model compound) and $Co^{III}(dmgH)_2pyCl$ (NAD^+ reduction catalyst) (Scheme 4.3) is virtually the same as the case of photocatalytic H_2 evolution from X-QH_2 with

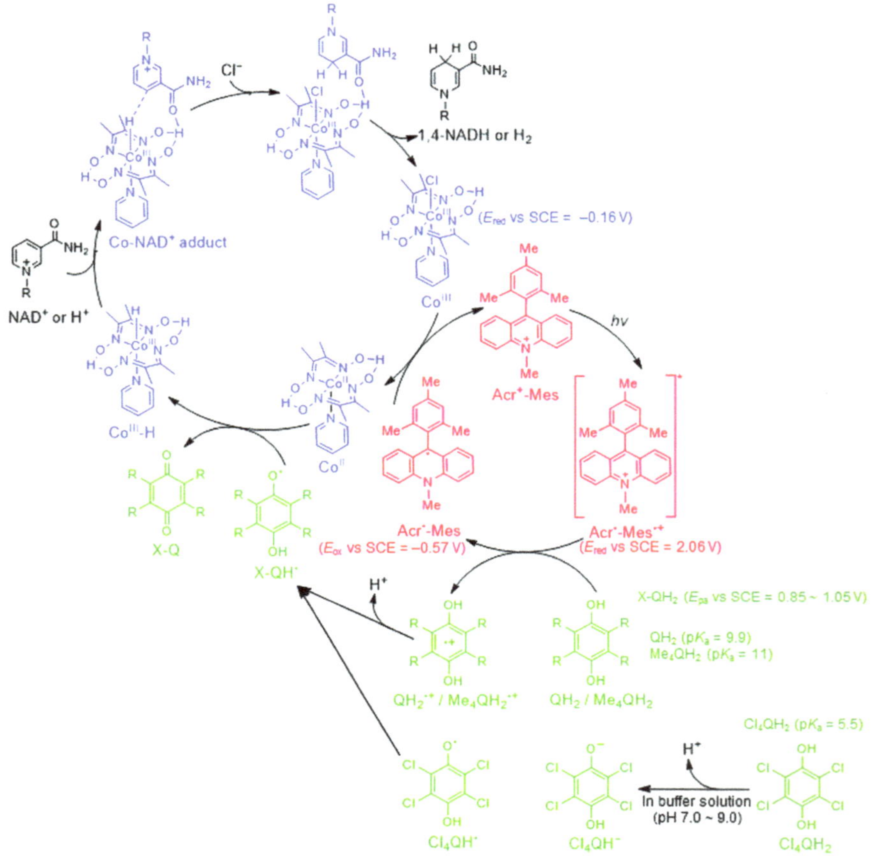

Scheme 4.3 Mechanism in the photocatalytic reduction of NAD^+ to NADH by X-QH_2 with Acr^+–Mes (a PRC model compound) and $Co^{III}(dmgH)_2pyCl$ (NAD^+ reduction catalyst). Reproduced from ref. 25 with permission from American Chemical Society, Copyright 2024.

Acr^+–Mes and $Co^{III}(dmgH)_2pyCl$ (Scheme 4.2).[25] Firstly, photoexcitation of Acr^+–Mes result is generation of the ET state $^3(Acr^•–Mes^{•+})$, which undergoes the ET oxidation of X-QH_2 to X-$QH_2^{•+}$ and the ET reduction of $Co^{III}(dmgH)_2pyCl$ to $[Co^{II}(dmgH)_2pyCl]^-$.[25] X-$QH_2^{•+}$ is deprotonated to produce a semiquinone radical (X-$QH^•$),[24] followed by hydrogen atom transfer from X-$QH^•$ to $[Co^{II}(dmgH)_2pyCl]^-$ to produce X-Q and the Co(III)-hydride complex ($[Co(H)(dmgH)_2pyCl]^-$).[25] Hydride transfer from $[Co(H)(dmgH)_2pyCl]^-$ to NAD^+ occurs *via* a six-membered ring transition state, in which the H-ligand interacts with the C4-position of NAD^+ to yield NADH regioselectively (1,4-dihydro form).[26,27] No other regioisomers such as 1,2- and 1,6-NADH were produced in the photocatalytic reduction of NAD^+ by X-QH_2 with Acr^+–Mes and $Co^{III}(dmgH)_2pyCl$ (Scheme 4.3).[25]

References

1. M. M. Najafpour and S. I. Allakhverdiev, *J. Photochem. Photobiol., B*, 2015, **152**, 173–175.
2. H. Kubota-Kawai, R. Mutoh, K. Shinmura, P. Setif, M. M. Nowaczyk, M. R. T. Ikegami, H. Tanaka and G. P. Kurisu, *Nat. Plants*, 2018, **4**, 218–224.
3. J. Li, N. Hamaoka, F. Makino, A. Kawamoto, Y. Lin, M. Rögner, M. M. Nowaczyk, Y.-H. Lee, K. Namba, C. Gerle and G. Kurisu, *Commun. Biol.*, 2022, **5**, 951.
4. B. D. Kossalbayev, G. Yilmaz, A. K. Sadvakasova, B. K. Zayadan, A. M. Belkozhayev, G. K. Kamshybayeva, G. A. Sainova, A. M. Bozieva, H. F. Alharby, T. Tomo and S. I. Allakhverdiev, *Int. J. Hydrogen Energy*, 2024, **49**, 413–432.
5. A. M. Bozieva, M. K. Khasimov, R. A. Voloshin, M. A. Sinetova, E. V. Kupriyanova, S. K. Zharmukhamedov, D. O. Dunikov, A. A. Tsygankov, T. Tomo and S. I. Allakhverdiev, *Int. J. Hydrogen Energy*, 2023, **48**, 7569–7581.
6. W. T. Eckenhoff, *Coord. Chem. Rev.*, 2018, **373**, 295–316.
7. K. E. Dalle, J. Warnan, J. J. Leung, B. Reuillard, I. S. Karmel and E. Reisner, *Chem. Rev.*, 2019, **119**, 2752–2875.
8. S. Fukuzumi, Y.-M. Lee and W. Nam, *Coord. Chem. Rev.*, 2018, **355**, 54–73.
9. L. Tong, L. Duan, A. Zhou and R. P. Thummel, *Coord. Chem. Rev.*, 2020, **402**, 213079.
10. F. Droghetti, F. Lucarini, A. Molinari, A. Ruggi and M. Natali, *Dalton Trans.*, 2022, **51**, 10658–10673.
11. J. S. O'Neill, L. Kearney, M. P. Brandon and M. T. Pryce, *Coord. Chem. Rev.*, 2022, **467**, 214599.
12. L. Leone, G. Sgueglia, S. La Gatta, M. Chino, F. Nastri and A. Lombardi, *Int. J. Mol. Sci.*, 2023, **24**, 8605.
13. J. Fortage, M.-N. Collomb and C. Costentin, *ChemSusChem*, 2024, **17**, e202400205.
14. S. J. Phukan, S. Goswami, S. Bhowmik, N. K. Sah, M. Sharma, P. Pramanik, C. Pathak, M. Roy, R. K. Pai and S. Gara, *Fuel*, 2024, **361**, 130654.
15. D. N. Tritton, F. K. Tang, G. B. Bodedla, F. W. Lee, C. S. Kwan, K. C.-F. Leung, X. Zhu and W.-Y. Wong, *Coord. Chem. Rev.*, 2022, **459**, 214390.
16. S. Fukuzumi, *Opin. Chem. Biol.*, 2015, **25**, 18–26.
17. A. Mazzeo, S. Santalla, C. Gaviglio, F. Doctorovich and J. Pellegrino, *Inorg. Chim. Acta*, 2021, **517**, 11950.
18. Y. H. Hong, Y.-M. Lee, W. Nam and S. Fukuzumi, *ACS Catal.*, 2023, **13**, 308–341.
19. S. Fukuzumi, K. Ohkubo and T. Suenobu, *Acc. Chem. Res.*, 2014, **47**, 1455–1464.
20. T. Tsudaka, H. Kotani, K. Ohkubo, T. Nakagawa, N. V. Tkachenko, H. Lemmetyinen and S. Fukuzumi, *Chem. – Eur. J.*, 2017, **23**, 1306–1317.
21. J. L. Dempsey, B. S. Brunschwig, J. R. Winkler and H. B. Gray, *Acc. Chem. Res.*, 2009, **4**, 1995–2004.
22. Y. H. Hong, Y.-M. Lee, W. Nam and S. Fukuzumi, *Inorg. Chem.*, 2020, **59**, 14838–14846.
23. K. Ohkubo, K. Mizushima, R. Iwata, K. Souma, N. Suzuki and S. Fukuzumi, *Chem. Commun.*, 2010, **46**, 601–603.
24. O. Brede, S. Kapoor, T. Mukherjee, R. Hermanna and S. Naumov, *Phys. Chem. Chem. Phys.*, 2002, **4**, 5096–5104.
25. Y. H. Hong, M. Nilajakar, Y.-M. Lee, W. Nam and S. Fukuzumi, *J. Am. Chem. Soc.*, 2024, **146**, 5152–5161.
26. J. A. Kim, S. Kim, J. Lee, J.-O. Baeg and J. Kim, *Inorg. Chem.*, 2012, **51**, 8057–8063.
27. S. Fukuzumi, Y.-M. Lee and W. Nam, *J. Inorg. Biochem.*, 2019, **199**, 110777.

5 PSII Models

5.1 OEC Models

In photosynthesis, photodriven water oxidation is catalysed by a $CaMn_4O_5$ cluster in the oxygen evolving complex (OEC) in PSII.[1–6] The X-ray crystal structure is shown in Figure 5.1.[7] The Ca^{2+} ion in the $CaMn_4O_5$ cluster plays an essential role in the four-electron/four-proton oxidation of H_2O to O_2.[4] Without Ca^{2+}, no water oxidation proceeded.[4]

Binding of a metal ion to the oxo moiety of high-valent manganese–oxo complexes, as observed by extended X-ray absorption fine structure (EXAFS) spectroscopy, indicated that going from no Sc^{3+} binding in a Mn(IV)–oxo complex, $[(N4Py)Mn^{IV}(O)]^{2+}$ (N4Py = *N*,*N*-bis(2-pyridylmethyl)-*N*-bis(2-pyridyl)methylamine), to one and two Sc^{3+} binding, results in elongation of the Mn=O bond from 1.69 to 1.74(2) Å, which is consistent with a weakening of the Mn=O bond.[8] The EXAFS data also revealed a short Mn–Sc distance (3.45(10) Å), which clearly indicates that Sc^{3+} ions bind to the $Mn^{IV}(O)$ moiety in both $[(N4Py)Mn^{IV}(O)]^{2+}$–$(Sc^{3+})_1$ and $[(N4Py)Mn^{IV}(O)]^{2+}$–$(Sc^{3+})_2$.[8] DFT calculations of the $Mn^{IV}(O)$–$(Sc^{3+})_2$ complex suggested that one Sc^{3+} ion binds directly to the oxo moiety of the $Mn^{IV}(O)$ complex but the second Sc^{3+} ion is located at the secondary coordination sphere (Figure 5.2).[8]

The E_{red} values of $Mn^{IV}(O)$ complexes, $[(Bn\text{-}TPEN)Mn^{IV}(O)]^{2+}$ [(Bn-TPEN = *N*-benzyl-*N*,*N*′,*N*′-tris(2-pyridylmethyl)-1,2-diaminoethane)] and $[(N4Py)Mn^{IV}(O)]^{2+}$, in the presence of various concentrations of $Sc(OTf)_3$, were determined from the ET equilibrium constant (K_{et})

RSC Foundations No. 1
Artificial Photosynthesis
By Shunichi Fukuzumi

Published by the Royal Society of Chemistry, www.rsc.org

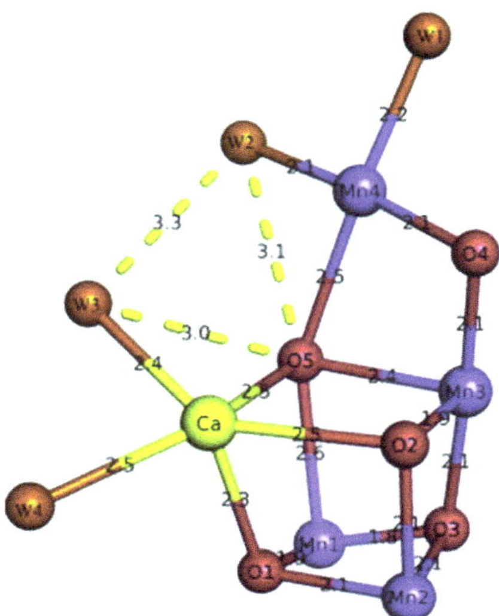

Figure 5.1 X-ray crystal structure of the oxygen evolving complex (OEC) in photosystem II (PSII). Reproduced from ref. 7 with permission from Elsevier, Copyright 2011.

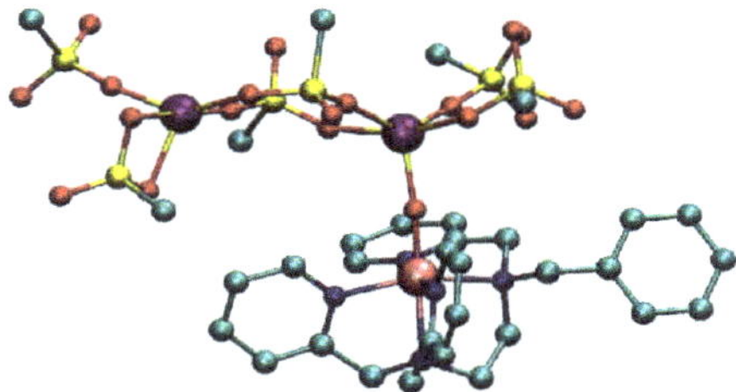

Figure 5.2 Optimized structure of $[(Bn\text{-}TPEN)Mn^{IV}(O)\text{-}[Sc(OTf)_3]_2]^{2+}$, calculated by DFT at the B3LYP/LACVP level. The Mn–O bond lengths of $[Mn^{IV}(O)]^{2+}$, $[Mn^{IV}(O)]^{2+}\text{-}Sc^{3+}$, and $[Mn^{IV}(O)]^{2+}\text{-}(Sc^{3+})_2$ were calculated to be 1.68, 1.75, and 1.75 Å, respectively. Reproduced from ref. 8 with permission from American Chemical Society, Copyright 2013.

values obtained by the ET titrations with $[Ru^{II}(bpy)_3]^{2+}$ (E_{ox} *vs.* SCE = 1.24 V) and $[Ru^{II}(5\text{-}Cl\text{-}phen)_3]^{2+}$ (5-Cl-phen = 5-chloro-1,10-phenanthroline; E_{ox} *vs.* SCE = 1.36 V).[9] The E_{red} values (*vs.* SCE) of $[(Bn\text{-}TPEN)Mn^{IV}(O)]^{2+}$ and $[(N4Py)Mn^{IV}(O)]^{2+}$ in the absence of $Sc(OTf)_3$ increased from 0.78 and 0.80 V to 1.36 and 1.42 V in the presence of large excess $Sc(OTf)_3$, which correspond to the E_{red} values of $[(Bn\text{-}TPEN)Mn^{IV}(O)]^{2+}\text{-}(Sc^{3+})_2$ and $[(N4Py)Mn^{IV}(O)]^{2+}\text{-}(Sc^{3+})_2$, respectively (Figure 5.3).[9] Such large shifts in the E_{red} values result from strong binding of Sc^{3+} ions to the oxo moiety of $[(Bn\text{-}TPEN)Mn^{IV}(O)]^{2+}$ and $[(N4Py)Mn^{IV}(O)]^{2+}$.[9] In the

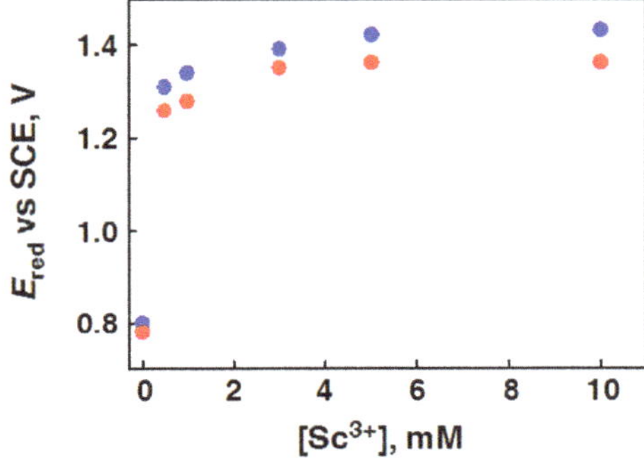

Figure 5.3 Plot of E_{red} (*vs.* SCE) of $[(N4Py)Mn^{IV}(O)]^{2+}$ (blue circles) and $[(Bn\text{-}TPEN)Mn^{IV}(O)]^{2+}$ (red circles) on [Sc^{3+}] in CF_3CH_2OH/CH_3CN (v/v = 1:1) at 273 K *vs.* concentration of Sc^{3+}. The E_{red} values were determined from the ET equilibrium constants (K_{et}) between electron donors and $[(N4Py)Mn^{IV}(O)]^{2+}$. Reproduced from ref. 9 with permission from American Chemical Society, Copyright 2013.

case of $[(TMC)Fe^{IV}(O)]^{2+}$, the X-ray crystal structure of the Sc^{3+}-bound $[(TMC)Fe^{IV}(O)]^{2+}$ complex was reported previously.[10] Later, Sc^{3+} was suggested to bind to the Fe(III)–oxo moiety.[11] In any case, this is the first case showing the binding of Sc^{3+} to the oxo moiety of the $Fe^{IV}(O)$ or $Fe^{III}(O)$ moiety.[10]

Two molecules of triflic acid (HOTf), which is a stronger Lewis acid than $Sc(OTf)_3$, can also bind to the oxo moiety of $[(N4Py)Mn^{IV}(O)]^{2+}$ to produce $[(N4Py)Mn^{IV}(O)]^{2+}$–$(HOTf)_2$, when ET from electron donors to $[(N4Py)Mn^{IV}(O)]^{2+}$–$(HOTf)_2$ proceeds *via* an outer-sphere pathway rather than an inner-sphere pathway because of the steric effects of two HOTf molecules bound to the oxo moiety of $[(N4Py)Mn^{IV}(O)]^{2+}$.[12] In such a case, the logarithm of the rate constants of epoxidation of styrene derivatives, sulfoxidation of thioanisole derivatives and hydrogen atom transfer reactions with $[(N4Py)Mn^{IV}(O)]^{2+}$–$(HOTf)_2$ as well as ET from coordinatively-saturated metal complexes such as ferrocene derivatives to $[(N4Py)Mn^{IV}(O)]^{2+}$–$(HOTf)_2$ shows a remarkably unified correlation with the ET driving force in light of the Marcus theory of outer-sphere ET (Figure 5.4).[13] Thus, mechanisms of oxygen atom transfer (epoxidation and sulfoxidation) reactions of styrene and thioanisole derivatives and hydrogen atom transfer (hydroxylation) reactions of toluene derivatives by $[(N4Py)Mn^{IV}(O)]^{2+}$–$(HOTf)_2$ have been unified as the same reaction pathway *via* outer-sphere ET, followed by the fast bond-forming step. In the case of the epoxidation of *cis*-stilbene by $[(N4Py)Mn^{IV}(O)]^{2+}$–$(HOTf)_2$, the isomerization of *cis*-stilbene radical cation to *trans*-stilbene radical cation occurs after

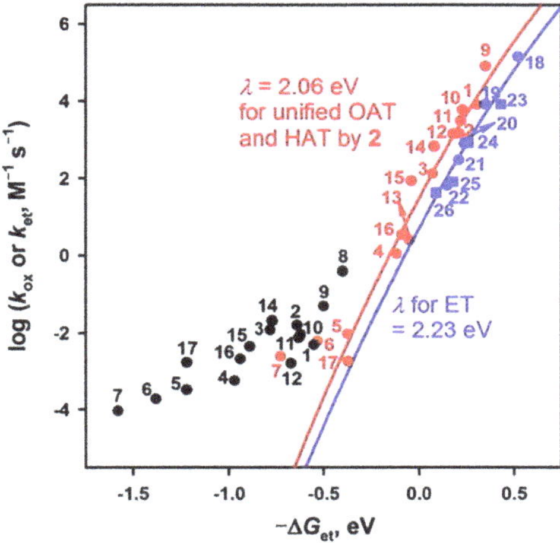

Figure 5.4 Plots of log k_{ox} for epoxidation of styrene derivatives [(1) 4-MeO-styrene, (2) *trans*-stilbene, (3) *cis*-stilbene, (4) 4-Me-styrene, (5) styrene, (6) 3-Cl-styrene and (7) 2,6-di-Cl-styrene], sulfoxidation of thioanisole derivatives [(8) 4-MeO-thioanisole, (9) 4-Me-thioanisole, (10) thioanisole, (11) 4-F-thioanisole, (12) 4-Br-thio-anisole and (13) 4-CN-thioanisole] and hydroxylation of toluene derivatives [(14) hexa-methylbenzene, (15) pentamethylbenzene, (16) 1,2,4,5-tetramethylbenzene and (17) mesitylene] by $[(N4Py)Mn^{IV}(O)]^{2}$ (black circles) and $[(N4Py)Mn^{IV}(O)]^{2+}-(HOTf)_2$ (red circles) and log k_{et} for ET from electron donors [coordinatively saturated metal complexes; (18) $[Ru^{II}(Me_2bpy)_3]^{2+}$, (19) $[Ru^{II}(bpy)_3]^{2+}$, (20) $[Ru^{II}(5\text{-}Clphen)_3]^{2+}$, (21) $[Ru^{II}(5\text{-}Br\text{-}bpy)_3]^{2+}$ and (22) $[Ru^{II}(5\text{-}NO_2phen)_3]^{2+}$] to $[(N4Py)Mn^{IV}(O)]^{2+}-(HOTf)_2$ (blue circles) in TFE/MeCN (1 : 1 v/v) at 273 K *vs.* the ET driving force ($-\mathrm{D}G_{et}$). The blue squares show the driving force dependence of the rate constants (log k_{et}) of ET from electron donors [(23) ferrocene, (24) bromoferrocene, (25) acetylferrocene and (26) dibromoferrocene] to $[(N4Py)Mn^{IV}(O)]^{2+}$ in TFE/MeCN (1 : 1 v/v) at 273 K. Reproduced from ref. 13 with permission from American Chemical Society, Copyright 2019.

outer-sphere ET from *cis*-stilbene to $[(N4Py)Mn^{IV}(O)]^{2+}-(HOTf)_2$ to yield *trans*-stilbene oxide selectively.[13] This is taken as evidence for the occurrence of outer-sphere ET in the acid-catalyzed epoxidation, because the *cis*-stilbene radical cation is rapidly isomerized to a *trans*-stilbene radical cation, which affords *trans*-stilbene oxide selectively.[13]

A similar unified driving force dependence of log k_{ox} was reported for oxygen atom transfer and ET reactions of $[(N4Py)Mn^{IV}(O)]^{2+}-(Sc(OTf)_3)_2$.[14] The logarithm of k_{ox} values of metal ion-bound Mn(IV)–oxo complexes increased linearly with increasing the Lewis acidity of metal ions (ΔE), which was determined from the g_{zz} values of $O_2^{\bullet -}-M^{n+}$ complexes (Figure 5.5).[16–18]

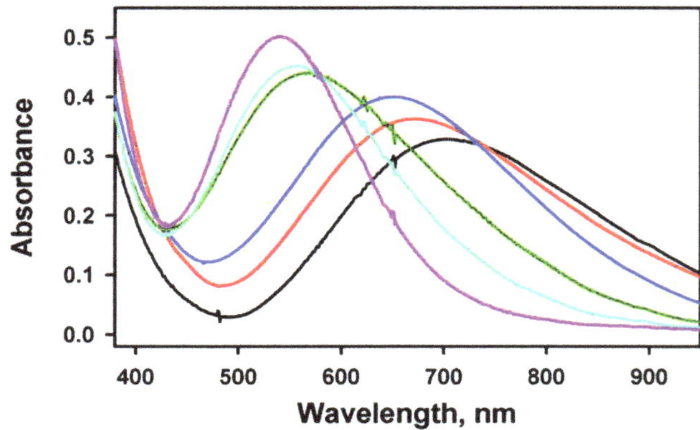

Figure 5.5 UV-vis spectra of $[(TMC)Fe^{III}(O_2)]-M^{n+}$ complexes $[M^{n+} = Sc^{3+}$ (pink), Y^{3+} (cyan), Lu^{3+} (green), Zn^{2+} (blue), Ca^{2+} (red) and Sr^{2+} (black)] obtained in the reaction of the isolated $[(TMC)Fe^{III}(O_2)](OTf)$ (black, 0.50 mM) with redox-inactive metal ions (M^{n+}) in MeCN at −20 °C: $\lambda_{max} = 535$ nm (Sc^{3+}), $\lambda_{max} = 560$ nm (Y^{3+}), $\lambda_{max} = 570$ nm (Lu^{3+}), $\lambda_{max} = 650$ nm (Zn^{2+}), $\lambda_{max} = 670$ nm (Ca^{2+}) and $\lambda_{max} = 710$ nm (Sr^{2+}). Reproduced from ref. 15 with permission from Springer Nature, Copyright 2014.

The reaction of $[(TMC)Fe^{II}]^{2+}$ with metal ion-bound superoxide ($O_2^{\bullet-}-M^{n+}$) afforded the iron(III)-peroxo-metal ion complex ($[(TMC)Fe^{III}(O_2)]^+-M^{n+}$).[15,19] The $[(TMC)Fe^{III}(\mu,\eta^2:\eta^2\text{-}O_2)]^+-M^{n+}$ complexes were also prepared by binding of redox-inactive metal ions to the peroxo moiety of $[(TMC)Fe^{III}(O_2)]^+$.[15,19] The energy of the absorption maxima ($h\nu_{max}$) due to the ligand to metal charge transfer (LMCT) transition of $[(TMC)Fe^{III}(O_2)]^+-M^{n+}$ varied depending on the Lewis acidity (ΔE) of metal ions and the $h\nu_{max}$ value increased linearly with increasing the Lewis acidity of M^{n+} (ΔE), which was also determined from the g_{zz} values of $O_2^{\bullet-}-M^{n+}$ complexes (Figure 5.6).[15–17] The $h\nu_{max}$ value of $[(TMC)Fe^{III}(O_2)]^+-Ca^{2+}$ is similar to that of $[(TMC)Fe^{III}(O_2)]^+-Sr^{2+}$.[19] The EXAFS data together with DFT calculations indicate that the calculated Fe–M^{2+} distances and the Fe–O–M^{2+} angles in $[(TMC)Fe^{III}(O_2)]^+-Ca^{2+}$ and $[(TMC)Fe^{III}(O_2)]^+-Sr^{2+}$ are also similar (4.11 Å and 4.34 Å and 140° and 142°, respectively).[15]

The ET reduction of $[(TMC)Fe^{III}(O_2)]^+-M^{n+}$ resulted in cleavage of the O–O bond, being accelerated by increasing the Lewis acidity of M^{n+}.[20] In sharp contrast, the ET oxidation of $[(TMC)Fe^{III}(O_2)]^+-M^{n+}$ resulted in the release of O_2 being decelerated by increasing the Lewis acidity of M^{n+}.[20] Thus, the Lewis acidity of metal ions to bind high-valent metal–oxo species must be strong enough to facilitate the O–O bond formation step by increasing the oxidizing capability of

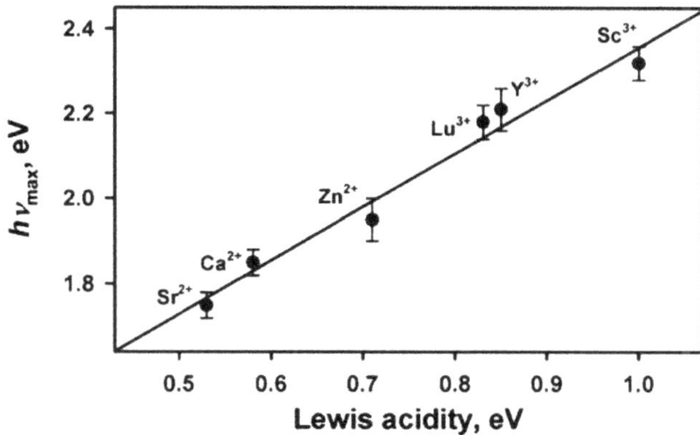

Figure 5.6 Plot of $h\nu_{max}$ of $[(TMC)Fe^{III}(O_2)]^+$–M^{n+} complexes *vs.* Lewis acidity of metal ions (ΔE). Reproduced from ref. 15 with permission from Springer Nature, Copyright 2014.

the metal–oxo species. However, the Lewis acidity of metal ions to bind metal-peroxo species must be weak enough not to retard the O_2 release by decreasing the reducing capability of metal-peroxo species. The Lewis acidity of Ca^{2+} and Sr^{2+} may be strong enough to facilitate the oxidizing ability of high-valent manganese–oxo species, but weak enough not to decelerate the reducing ability of manganese–peroxo species to release O_2 in the OEC.[15]

This may be a reason why nature did not choose a stronger Lewis acidic redox-inactive metal ion (*e.g.*, Zn^{2+}) but instead chose a Ca^{2+} ion in the oxidation of water to evolve O_2 and why Sr^{2+} ion, which has virtually the same Lewis acidity as Ca^{2+}, is the only surrogate that replaces the Ca^{2+} ion for the reactivity of the OEC in PSII.[15]

In order to stabilize water oxidation catalysts, metal complexes of inorganic ligands, which are stable under water oxidation conditions, are suitable for catalytic water oxidation. A tetraruthenium complex with silicodecatungstate, $[\{Ru_4O_4(OH)_2(H_2O)_4\}(\gamma\text{-}SiW_{10}O_{36})_2]^{10-}$, was reported to act as a robust water oxidation catalyst with one-electron oxidants such as diammonium cerium(IV) nitrate, $(NH_4)_2[Ce^{IV}(NO_3)_6]$ (CAN), and $[Ru(bpy)_3]^{3+}$ (bpy = 2,2′-bipyridine).[21] A tetracobalt complex with phosphononatungstate $[Co_4(H_2O)_2(PW_9O_{34})_2]^{10-}$ was also reported to be capable of oxidizing water with $[Ru^{III}(bpy)_3]^{3+}$.[22] Detection of a high-valent metal–oxo complex during the catalytic water oxidation was reported in the catalytic oxidation of water by CAN with mononuclear Ru complexes bearing a Keggin-type lacunary heteropolytungstate, $[Ru^{III}(H_2O)SiW_{11}O_{39}]^{5-}$ and $[Ru^{III}(H_2O)GeW_{11}O_{39}]^{5-}$

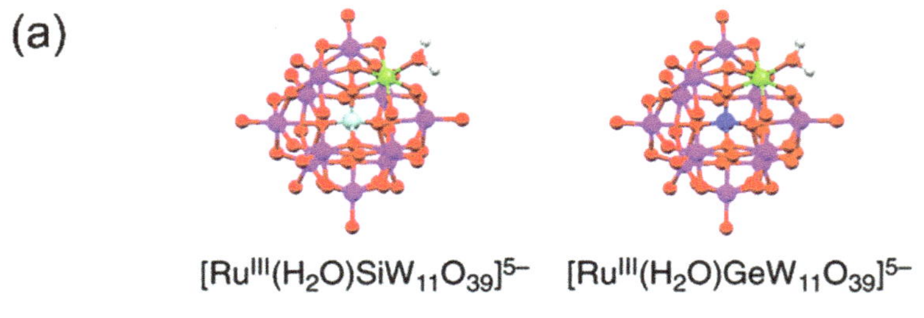

Figure 5.7 (a) Structures of polyanions, $[(L)Ru^{III}(H_2O)]^{5-}$ ($L = SiW_{11}O_{39}$ and $GeW_{11}O_{39}$; Ru: light green, W: purple, O: red, H: white, Si: light blue, Ge: deep blue). (b) Proposed mechanism of the catalytic water oxidation by Ce^{IV} with $[(L)Ru^{III}(H_2O)]^{5-}$ ($L = SiW_{11}O_{39}$ and $GeW_{11}O_{39}$). Reproduced from ref. 23 with permission from American Chemical Society, Copyright 2011.

(Figure 5.7a).[23] The stepwise ET from $Ru^{III}–OH_2$ complexes to CAN, followed by deprotonation of two protons from the aqua ligand afforded the $Ru^{V}{=}O$ complexes (Figure 5.7b).[23] The $Ru^{V}{=}O$ complex derived from $[Ru^{III}(H_2O)SiW_{11}O_{39}]^{5-}$ was detected by UV-vis absorption, EPR and resonance Raman measurements *in situ* as active species during the catalytic oxidation of water by CAN.[23] The rate of the catalytic oxidation of water by excess CAN with $[Ru^{III}(H_2O)SiW_{11}O_{39}]^{5-}$ and $[Ru^{III}(H_2O)GeW_{11}O_{39}]^{5-}$ in an acidic medium obeyed first-order kinetics with respect to catalyst concentration, suggesting that the reaction of a $Ru^{V}{=}O$ intermediate with water to form a $Ru^{III}–OOH$ species is the rate-determining step for the O–O bond formation (Figure 5.7b).[23] The rate of water oxidation increased to reach a constant value with increasing pH and concentration of CAN.[23] Such a saturation behaviour suggests that the subsequent ET oxidation of the $Ru^{III}–OOH$ species by CAN, which results in release of O_2, competes with the reverse reaction from the $Ru^{III}–OOH$ complex, which undergoes O–O bond cleavage to regenerate the $Ru^{V}{=}O$ complex (Figure 5.7b).[23]

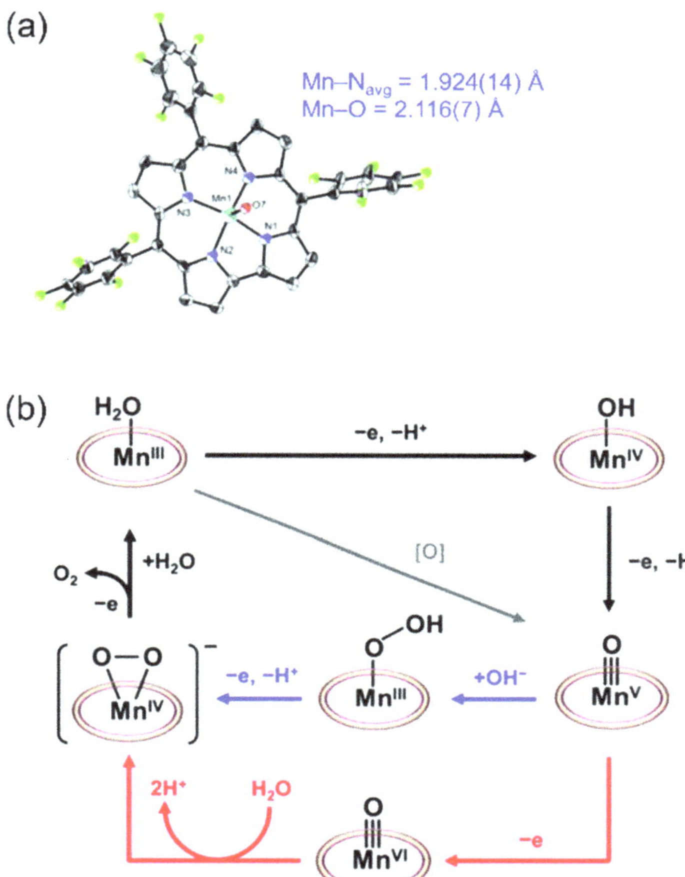

Figure 5.8 (a) Molecular structure of Mn^{III}{tris(pentafluorophenyl)corrole}. (b) A proposed catalytic mechanism of water oxidation with the Mn^{III} corrole complex. Reproduced from ref. 24 with permission from American Chemical Society, Copyright 2021.

Detection of a Mn^{IV}–peroxo intermediate was reported by Cao and coworkers in the electrocatalytic oxidation of water with Mn^{III}{tris(pentafluorophenyl)corrole} (Figure 5.8a) in propylene carbonate (PC) *via* nucleophilic attack of H_2O on a high-valent terminal Mn–oxo species.[24] A Mn^{IV}–peroxo species was identified as an intermediate by various spectroscopic methods, including UV-vis, electron paramagnetic resonance (EPR) and infrared (IR) spectroscopies, during the electrolysis of Mn^{III}{tris(pentafluorophenyl)-corrole} with water.[24] When OH^- was added to a solution of a Mn^V(O) species, the absorption band due to the Mn^V(O) complex rapidly disappeared within 1 second at room temperature.[24] The isotope labelling experiments showed that one O atom of evolved O_2 came from the $Mn^V(O)$ complex and the other originated from OH^-, suggesting the nucleophilic attack of OH^- on the $Mn^V(O)$ complex.[24] Density functional theory calculations also suggested that the nucleophilic attack of hydroxide on the $Mn^V(O)$ complex and the water nucleophilic attack on the one-electron-oxidized $Mn^V(O)$ were involved in the electrocatalytic cycle of water oxidation.[24]

Detection of a Mn^{IV}–peroxo intermediate led to proposing the catalytic mechanism of water oxidation with Mn^{III}{tris(pentafluorophenyl)-corrole}, as shown in Figure 5.8b, where the $Mn^V(O)$ complex is formed by electrochemical and chemical oxidation.[24] The nucleophilic attack of OH^- on $Mn^V(O)$ affords Mn^{III}–OOH, which is oxidized by unreacted $Mn^V(O)$ and deprotonated with excess of OH^- to produce the Mn^{IV}–peroxo species.[24] The exact structure of this Mn^{IV}–peroxo species, whether this is end-on or side-on, has yet to be clarified.[24] Because the structure of the side-on Mn^{IV}–peroxo species bearing a tetraamido macrocyclic ligand was determined by single crystal X-ray diffraction,[25] a side-on Mn^{IV}–peroxo structure was proposed, as shown in Figure 5.8b.[24] The nucleophilic attack of water on the $Mn^V(O)$ complex was much slower than that of OH^-.[24] Thus, further oxidation of the $Mn^V(O)$ complex is required to react with H_2O under electrocatalytic conditions.[24] It was proposed that the $Mn^{VI}(O)$ species was formed by the ET oxidation of the $Mn^V(O)$ complex, based on DFT calculations.[24] The nucleophilic attack of water on the $Mn^{VI}(O)$ species, accompanied by deprotonation, produces the Mn^{IV}–peroxo species, which was further oxidized to evolve O_2, accompanied by regeneration of the Mn^{III} complex.[24]

Photocatalytic oxidation of water by persulfate with a homogeneous catalyst was also reported by Dhar's group using Fe^{III} complexes of biuret-modified tetraamido macrocyclic ligands (b-TAML) in the presence of $[Ru^{II}(bpy)_3]^{2+}$ as a photoredox catalyst with TON of 220 and TOF of 0.76 s^{-1} in a pH 8.7 borate buffer.[26] A high-valent $Fe^V(O)$ intermediate was suggested to be formed in the photocatalytic oxidation of H_2O to O_2 with $[Fe^{III}(b\text{-TAML})]^-$ as identified using EPR, UV-vis and resolution mass spectrometry (HRMS) studies.[26] It should be noted that the X-ray crystal structure of $[Fe^V(O)(TAML)]^-$ was reported earlier by Corrin's group.[27] It was proposed that the nucleophilic attack of water on the $Fe^V(O)$ complex resulted in the formation of Fe^{III}–hydroperoxo species, which was further oxidized to release O_2.[26] Photoirradiation of a buffer solution containing $[Fe^{III}(b\text{-TAML})]^-$, $Na_2S_2O_8$ and $[Ru^{II}(bpy)_3]^{2+}$ resulted in ET from $[Ru(bpy)_3]^{2+*}$ to $S_2O_8^{2-}$

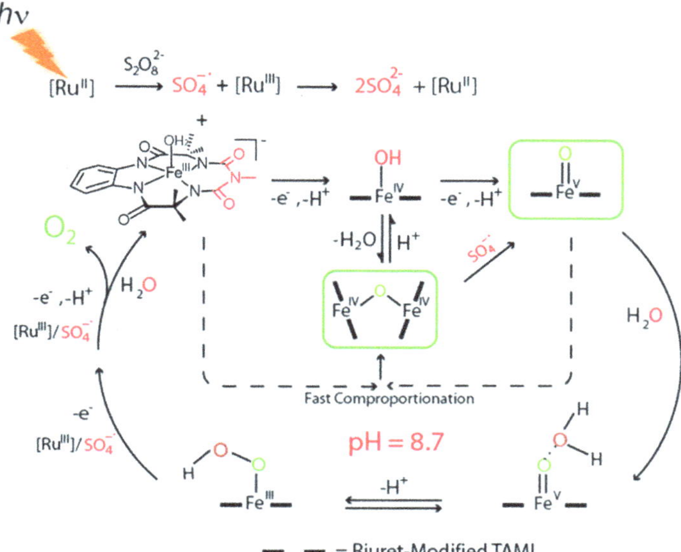

Scheme 5.1 A proposed photocatalytic mechanism of water oxidation by $[Ru^{III}(bpy)_3]^{3+}$ with Fe(b-TAML). Reproduced from ref. 26 with permission from American Chemical Society, Copyright 2014.

to produce $[Ru(bpy)_3]^{3+}$, $SO_4^{\bullet-}$ and SO_4^{2-}, as shown in Scheme 5.1, where $[Fe^{III}(H_2O)(b\text{-TAML})]^-$ is oxidized by $[Ru(bpy)_3]^{3+}$or $SO_4^{\bullet-}$ to produce the $Fe^{IV}(OH)$ complex, followed by the generation of an $Fe^V(O)$ complex *via* a proton-coupled ET (PCET) process (Scheme 5.1).[26] The $Fe^{IV}(OH)$ complex is converted to the μ-oxo dimer complex, which may also be produced by the comproportionation reaction of the $Fe^V(O)$ complex with the Fe^{III} complex.[26] The $Fe^V(O)$ reactive intermediate reacts with H_2O, followed by deprotonation to produce the $Fe^{III}(OOH)$ complex, which is further oxidized by two equivalents of $[Ru(bpy)_3]^{3+}$ or $SO_4^{\bullet-}$ to release O_2 (Scheme 5.1).[26]

5.2 Photocatalytic Water Oxidation by PQ Models

The oxidizing equivalents produced at the donor side of PSII are used to oxidize H_2O to O_2 (four electron/four proton oxidation), whereas the reducing equivalents accumulated at the acceptor side of PSII are used to reduce two ubiquinone molecules, Q_A and Q_B, which act as one-electron and two-electron gates, respectively.[1] Electrons and protons are transferred to plastoquinone (PQ) in quinone pool to produce two equivalents of plastoquinol (PQH_2).[28] The overall oxidation of H_2O to O_2, accompanied by reduction of PQ to PQH_2, is given by eqn (5.1), where two equivalents of PQ are reduced by two equivalents of H_2O to produce one equivalent of O_2 and two equivalents of PQH_2.[8]

$$2H_2O + 2PQ \xrightarrow[\text{PSII}]{h\nu} O_2 + 2PQH_2 \tag{5.1}$$

Photocatalytic oxidation of water to evolve O_2 has been studied extensively using sacrificial electron donors such as $Na_2S_2O_8$.[29–35] Photochemical four-electron/four-proton oxidation of H_2O, accompanied by the two-electron/two-proton reduction of *p*-benzoquinone derivatives (X-Q) used as PQ analogues in eqn (5.1) , which is a molecular functional model of PSII, has been made possible by using a nonheme iron(II) complex, $[(N4Py)Fe^{II}]^{2+}$ (N4Py = *N*,*N*-bis(2-pyridylmethyl)-*N*-bis(2-pyridyl)methylamine) as a water oxidation catalyst [eqn (5.2)].[36]

$$2H_2O + 2X\text{-}Q \xrightarrow[{[(N4Py)Fe^{II}]^{2+}}]{h\nu} O_2 + 2X\text{-}QH_2 \tag{5.2}$$

Photoirradiation of an MeCN solution of 2,3-dichloro-5,6-dicyano-*p*-benzoquinone (DDQ) containing $[(N4Py)Fe^{II}]^{2+}$ and H_2O (0.50 M) with a Xenon lamp resulted in O_2 evolution, as shown in Figure 5.9, where the O_2 yield reached nearly 100% based on the initial amount of DDQ.[36] Accompanied by H_2O oxidation to O_2, DDQ was reduced by H_2O to produce $DDQH_2$ [eqn (5.2), where X-Q is replaced by DDQ].[36] At prolonged illumination time, the O_2 yield decreased because of the photochemical oxidation of $DDQH_2$ by O_2 to produce H_2O_2, accompanied by regeneration of DDQ [eqn (5.3)].[36] Isotope-labelling experiments using ^{18}O-enriched water (97.4%) confirmed that the evolved $^{36}O_2$ (99% based on the labelled $H_2{}^{18}O$ percent) came from $H_2{}^{18}O$.[36] When DDQ was replaced by *p*-benzoquinone (Q), *p*-chloranil (CA) and 2,5-dimethyl-*p*-benzoquinone (PXQ), the O_2 yields were 90%.[36] In the case of tetramethyl-*p*-benzoquinone (DQ), however, the O_2 yield became smaller (*ca.* 50%).[36]

$$DDQH_2 + O_2 \xrightarrow{h\nu} DDQ + H_2O_2 \tag{5.3}$$

The catalytic mechanism of the photochemical oxidation of water by DDQ (that can be replaced by X-Q) with $[(N4Py)Fe^{II}]^{2+}$ is proposed based on detection of reaction intermediates (*vide infra*), as shown in Scheme 5.2.[36] Photoexcitation of DDQ results in the formation of $^3DDQ^*$ *via* intersystem crossing (ISC) from the singlet excited state

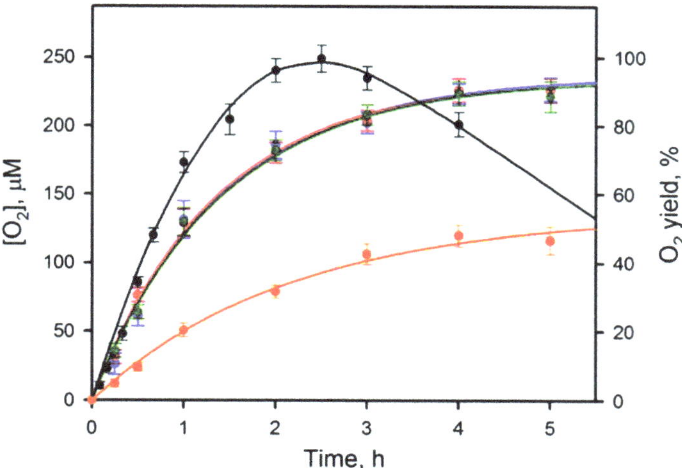

Figure 5.9 Time profiles of O_2 evolution in photochemical water oxidation by X-Q [0.50 mM; DDQ (black), BQ (blue), CA (green), DQ (orange), and PXQ (red)] with $[(N4Py)Fe^{II}]^{2+}$ (0.20 mM) under photoirradiation (white light) in an Ar-saturated MeCN solution containing H_2O (0.50 M) at 298 K. Reproduced from ref. 36 with permission from American Chemical Society, Copyright 2019.

Scheme 5.2 Proposed mechanism of photodriven water oxidation by DDQ with $[(N4Py)Fe^{II}]^{2+}$. Reproduced from ref. 36 with permission from American Chemical Society, Copyright 2019.

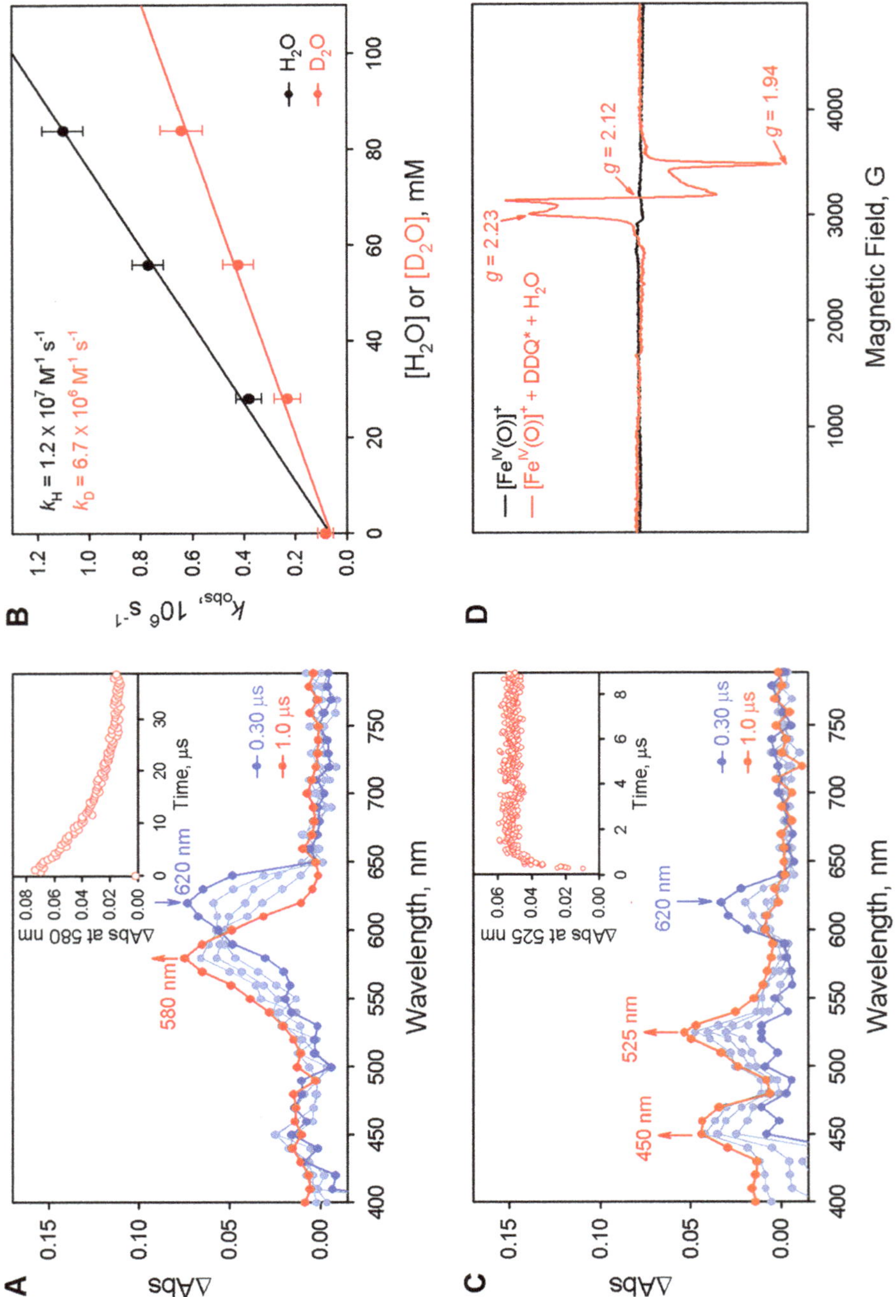
A
ΔAbs
580 nm
620 nm
ΔAbs at 580 nm
Time, μs
0.30 μs
1.0 μs
Wavelength, nm
B
k_{obs}, 10^6 s^{-1}
k_H = 1.2 X 10^7 M^{-1} s^{-1}
k_D = 6.7 X 10^6 M^{-1} s^{-1}
H_2O
D_2O
[H_2O] or [D_2O], mM
C
ΔAbs
450 nm
525 nm
620 nm
ΔAbs at 525 nm
Time, μs
0.30 μs
1.0 μs
Wavelength, nm
D
[$Fe^{IV}(O)]^+$
[$Fe^{IV}(O)]^+$ + DDQ* + H_2O
g = 2.23
g = 2.12
g = 1.94
Magnetic Field, G

($^1DDQ^*$).[37,38] ET from $[(N4Py)Fe^{II}]^{2+}$ to $^3DDQ^*$ occurs to produce $[(N4Py)Fe^{III}]^{3+}$ and $DDQ^{\bullet -}$ with a rate constant of 1.0×10^{10} M^{-1} s^{-1} in deaerated MeCN containing H_2O (0.50 M) at 298 K.[36] $[(N4Py)Fe^{III}]^{3+}$ reacts with water to produce the Fe(III)–hydroxo complex, $[(N4Py)Fe^{III}(OH)]^{2+}$, with release of H^+ which protonates $DDQ^{\bullet -}$ to afford $DDQH^{\bullet}$ (reaction pathway a in Scheme 5.2).[36] ET from $[(N4Py)Fe^{III}(OH)]^{2+}$ to $^3DDQ^*$ also proceeds to produce the Fe(IV)–oxo complex, $[(N4Py)Fe^{IV}(O)]^{2+}$ by release of H^+ which protonates $DDQ^{\bullet -}$ to afford $DDQH^{\bullet}$ (reaction pathway b in Scheme 5.2).[36] The formation of $[(N4Py)Fe^{IV}(O)]^{2+}$ was confirmed by the diagnostic visible absorption spectrum ($\lambda_{max} = 690$ nm) and CSI-MS.[39] $[(N4Py)Fe^{IV}(O)]^{2+}$ can be further oxidized to $[(N4Py)Fe^{V}(O)]^{2+}$ by ET from $[(N4Py)Fe^{IV}(O)]^{2+}$ to $^3DDQ^*$ with the rate constant of 9.4×10^9 M^{-1} s^{-1} at 298 K (reaction pathway c in Scheme 5.2).[36] Nucleophilic attack of water to the oxo moiety of $[(N4Py)Fe^{V}(O)]^{3+}$ proceeds rapidly to produce the Fe(III)–hydroperoxo complex ($[(N4Py)Fe^{III}(OOH)]^{2+}$) with release of H^+ which protonates $DDQ^{\bullet -}$ to afford $DDQH^{\bullet}$ (reaction pathway d in Scheme 5.2).[36] $DDQH^{\bullet}$ disproportionates to yield DDQ and $DDQH_2$.[36] $[(N4Py)Fe^{III}(OOH)]^{2+}$ may be thermally oxidized by DDQ *via* hydrogen atom transfer to afford $[(N4Py)Fe^{III}(O_2^{\bullet -})]^{2+}$ (reaction pathway e in Scheme 5.2), from which O_2 is released to regenerate $[(N4Py)Fe^{II}]^{2+}$ (reaction pathway f in Scheme 5.2).[36] It was confirmed that $[(N4Py)Fe^{III}(OOH)]^{2+}$, which was independently prepared by the reaction of $[(N4Py)Fe^{II}]^{2+}$ and H_2O_2,[40] was readily oxidized by DDQ thermally to yield O_2 with 100% yield based on DDQ concentration.[36]

Figure 5.10 (A) Transient absorption spectra observed upon laser excitation at 355 nm of a deaerated acetone solution of DDQ (0.40 mM) and $[Fe^{IV}(dpaq)(O)]^+$ (0.50 mM) at 0.30 μs (blue circles) and 1.0 μs (red circles) at 298 K. Inset shows the time course monitored at 580 nm due to $[Fe^{V}(dpaq)(O)]^{2+}$. (B) Plot of the first-order rate constant (k_{obs}) of decay of the absorption band at 580 nm due to $[Fe^{V}(dpaq)(O)]^{2+}$ *vs.* concentration of H_2O (black line) and D_2O (red line). (C) Transient absorption spectra observed upon laser excitation at 355 nm of a deaerated acetone solution of DDQ (0.40 mM), $[Fe^{IV}(dpaq)(O)]^+$ (0.50 mM) and H_2O (1.4 M) at 0.30 μs (blue circles) and 1.0 μs (red circles)] at 298 K. Inset shows the time course monitored at 525 nm due to the formation of $[Fe^{III}(dpaq)(OOH)]^+$. (D) EPR spectra of a deaerated acetone solution containing $[Fe^{IV}(dpaq)(O)]^+$ (black line) and a frozen acetone solution containing $[Fe^{IV}(dpaq)(O)]^+$ (0.50 mM) and DDQ (0.40 mM) in the presence of H_2O (150 mM) under photoirradiation for 1 min to produce $[Fe^{III}(dpaq)(OOH)]^+$ (red line). Spectrum was recorded at 5 K. Reproduced from ref. 41 with permission from Elsevier, Copyright 2024.

When $[Fe^{II}(N4Py)]^{2+}$ was replaced by $[Fe^{II}(dpaq)]^{+}$ (dpaq = 2-[bis(pyridine-2-ylmethyl)]amino-*N*-quinolin-8-yl-acetamido), the formation of $[Fe^{V}(dpaq)(O)]^{2+}$ (or $[Fe^{IV}$(dpaq radical cation)$(O)]^{2+}$) was directly detected using laser-induced transient absorption spectroscopy in the ET oxidation of $[Fe^{IV}(dpaq)(O)]^{+}$ by $^{3}DDQ^{*}$ (*vide infra*).[41] The formation of $[Fe^{III}(dpaq)(OOH)]^{+}$ was also directly observed by reacting $[Fe^{IV}(dpaq)(O)]^{+}$ with $[Ru^{III}(bpy)_3]^{3+}$ or $^{3}DDQ^{*}$ in the presence of water in acetone.[41] $[Fe^{IV}(dpaq)(O)]^{+}$ (0.50 mM) was independently prepared by the reaction of $[Fe^{II}(dpaq)]^{+}$ (0.50 mM) with O_2.[42]

Laser photoexcitation at 355 nm of a deaerated acetone solution of DDQ (0.40 mM) and $[Fe^{IV}(dpaq)(O)]^{+}$ (0.50 mM) resulted in the decay of absorbance at 620 nm due to $^{3}DDQ^{*}$, accompanied by the appearance of new absorption at 580 nm due to $[Fe^{V}(dpaq)(O)]^{2+}$ (Figure 5.10A).[41] The transient absorption band of $[Fe^{V}(dpaq)(O)]^{2+}$ ($\lambda_{max} = 580$ nm) was overlapped with that of $DDQ^{\bullet -}$ ($\lambda_{max} = 550$ and 590 nm).[41] This indicates that ET from $[Fe^{IV}(dpaq)(O)]^{+}$ to $^{3}DDQ^{*}$ occurred to produce $[Fe^{V}(dpaq)(O)]^{2+}$ and $DDQ^{\bullet -}$.[41]

The rate constant of ET from $[Fe^{IV}(dpaq)(O)]^{+}$ to $^{3}DDQ^{*}$ in the absence of water was determined from the slope of the plot of the first-order decay rate constant (k_{obs}) *vs.* concentration of $[Fe^{IV}(dpaq)(O)]^{+}$ to be $1.0(1) \times 10^{10}$ M^{-1} s^{-1} in acetone at 298 K.[41] In the presence of H_2O (1.4 M), the decay of absorbance at 620 nm due to $^{3}DDQ^{*}$ was accompanied by the appearance of a new absorption band at 525 nm due to $[Fe^{III}(dpaq)(OOH)]^{+}$ and that at 450 nm due to $DDQH^{\bullet}$ (Figure 5.10C).[41] This indicates that ET from $[Fe^{IV}(dpaq)(O)]^{+}$ to $^{3}DDQ^{*}$ in the presence of H_2O to produce $[Fe^{V}(dpaq)(O)]^{2+}$ and $DDQ^{\bullet -}$ is followed by the reaction of $[Fe^{V}(dpaq)(O)]^{2+}$ with H_2O to produce $[Fe^{III}(dpaq)(OOH)]^{+}$ and H^{+} which protonates $DDQ^{\bullet -}$ to afford $DDQH^{\bullet}$.[41] The rate constant of the reaction of $[Fe^{V}(dpaq)(O)]^{2+}$ with H_2O was also determined from the slope of the plot of k_{obs} *vs.* concentration of H_2O to be $1.2(1) \times 10^{7}$ M^{-1} s^{-1} in acetone at 298 K (Figure 5.10B).[41] When H_2O was replaced by D_2O, the rate constant of the reaction of $[Fe^{V}(dpaq)(O)]^{2+}$ with D_2O was determined to be $1.2(1) \times 10^{7}$ M^{-1} s^{-1}, which was smaller than that with H_2O in acetone at 298 K, affording the deuterium kinetic isotope effect (KIE) value of 1.8 (Figure 5.10B).[41] The KIE may result from the nucleophilic attack of water on the oxo moiety of $[Fe^{V}(dpaq)(O)]^{2+}$ in the O–O bond formation step, which is accompanied by the O–H bond cleavage of water.[41] Formation of $[Fe^{III}(dpaq)(OOH)]^{+}$ was confirmed by EPR signals observed at $g = 2.23$, 2.12, and 1.94 in a frozen solution containing $[Fe^{IV}(dpaq)(O)]^{+}$, DDQ, and H_2O under photoirradiation for 1 min

(Figure 5.10C).[41] These spectroscopic data agree with those of $[Fe^{III}(dpaq)(OOH)]^+$, which was produced in the reaction of $[Fe^{II}(dpaq)]^+$ with H_2O_2 in deaerated MeOH at 233 K.[42]

References

1. M. M. Najafpour, G. Renger, M. Hołyńska, A. N. Moghaddam, E.-M. Aro, R. Carpentier, H. Nishihara, J. J. Eaton-Rye, J.-R. Shen and S. I. Allakhverdiev, *Chem. Rev.*, 2016, **116**, 2886–2936.
2. J. Yano and V. Yachandra, *Chem. Rev.*, 2014, **114**, 4175–4205.
3. W. Lubitz, M. Chrysina and N. Cox, *Photosynth. Res.*, 2019, **142**, 105–125.
4. B. Semin, A. Loktyushkin and E. Lovyagina, *Biophys. Rev.*, 2024, **16**, 237–247.
5. M. M. Najafpour, I. Zaharieva, Z. Zand, S. M. Hosseini, M. Kouzmanova, M. Hołyńska, I. Tranca, A. W. Larkum, J.-R. Shen and S. I. Allakhverdiev, *Coord. Chem. Rev.*, 2020, **409**, 213183.
6. K. Yamaguchi, M. Shoji, H. Isobe, T. Kawakami, K. Miyagawa, M. Suga, F. Akita and J.-R. Shen, *Coord. Chem. Rev.*, 2022, **471**, 214742.
7. K. Kawakami, Y. Umena, N. Kamiya and J.-R. Shen, *J. Photochem. Photobiol., B*, 2011, **104**, 9–18.
8. J. Chen, X. Wu, K. M. Davis, Y.-M. Lee, M. S. Seo, K.-B. Cho, H. Yoon, Y. J. Park, S. Fukuzumi, Y. N. Pushkar and W. Nam, *J. Am. Chem. Soc.*, 2013, **135**, 6388–6391.
9. H. Yoon, Y.-M. Lee, X. Wu, K.-B. Cho, Y. N. Pushkar, W. Nam and S. Fukuzumi, *J. Am. Chem. Soc.*, 2013, **135**, 9186–9194.
10. S. Fukuzumi, Y. Morimoto, H. Kotani, P. Naumov, Y.-M. Lee and W. Nam, *Nat. Chem.*, 2010, **2**, 756–759.
11. J. Prakash, G. T. Rohde, K. K. Meier, A. J. Jasniewski, K. M. Van Heuvelen, E. Münck and L. Que Jr, *J. Am. Chem. Soc.*, 2015, **137**, 3478–3481.
12. J. Chen, H. Yoon, Y.-M. Lee, R. Sarangi, S. Fukuzumi and W. Nam, *Chem. Sci.*, 2015, **6**, 3624–3632.
13. Y.-M. Lee, S. Kim, K. Ohkubo, K.-H. Kim, W. Nam and S. Fukuzumi, *J. Am. Chem. Soc.*, 2019, **141**, 2614–2622.
14. N. Sharma, Y.-M. Lee, W. Nam and S. Fukuzumi, *Inorg. Chem.*, 2019, **58**, 13761–13765.
15. S. Bang, Y.-M. Lee, S. Hong, Y. Nishida, M. S. Seo, R. Sarangi, S. Fukuzumi and W. Nam, *Nat. Chem.*, 2014, **6**, 934–940.
16. S. Fukuzumi and K. Ohkubo, *Chem. – Eur. J.*, 2000, **6**, 4532–4535.
17. S. Fukuzumi and K. Ohkubo, *J. Am. Chem. Soc.*, 2002, **124**, 10270–10271.
18. D. G. Karmalkar, M. Sankaralingam, M. S. Seo, R. Ezhov, Y.-M. Lee, Y. N. Pushkar, W.-S. Kim, S. Fukuzumi and W. Nam, *Angew. Chem., Int. Ed.*, 2019, **58**, 16124–16129.
19. Y.-M. Lee, S. Bang, Y. M. Kim, J. Cho, S. Hong, T. Nomura, T. Ogura, O. Troeppner, I. Ivanović-Burmazović, R. Sarangi, S. Fukuzumi and W. Nam, *Chem. Sci.*, 2013, **4**, 3917–3923.
20. S. H. Bae, Y.-M. Lee, S. Fukuzumi and W. Nam, *Angew. Chem., Int. Ed.*, 2017, **56**, 801–805.
21. A. Sartorel, M. Carraro, G. Scorrano, R. D. Zorzi, S. Geremia, N. D. McDaniel, S. Bernhard and M. Bonchio, *J. Am. Chem. Soc.*, 2008, **130**, 5006–5007.
22. Q. Yin, J. M. Tan, C. Besson, Y. V. Geletii, D. G. Musaev, A. E. Kuznetsov, Z. Luo, K. I. Hardcastle and C. L. Hill, *Science*, 2010, **328**, 342–345.
23. M. Murakami, D. Hong, T. Suenobu and S. Fukuzumi, *J. Am. Chem. Soc.*, 2011, **133**, 11605–11613.
24. X. Li, X.-P. Zhang, M. Guo, B. Lv, K. Guo, X. Jin, W. Zhang, Y.-M. Lee, S. Fukuzumi, W. Nam and R. Cao, *J. Am. Chem. Soc.*, 2021, **143**, 14613–14621.

25. S. Hong, K. D. Sutherlin, J. Park, E. Kwon, M. A. Siegler, E. I. Solomon and W. Nam, *Nat. Commun.*, 2014, **5**, 5440.
26. C. Panda, J. Debgupta, D. D. Díaz, K. K. Singh, S. Sen Gupta and B. B. Dhar, *J. Am. Chem. Soc.*, 2014, **136**, 12273–12282.
27. F. T. iago de Oliveira, A. Chanda, D. Banerjee, X. Shan, S. Mondal and L. Que Jr, *Science*, 2007, **315**, 835–838.
28. F. Müh, C. Glöckner, J. Hellmich and A. Zouni, *Biochim. Biophys. Acta, Bioenerg.*, 2012, **1817**, 44–65.
29. J. D. Blakemore, R. H. Crabtree and G. W. Brudvig, *Chem. Rev.*, 2015, **115**, 12974–13005.
30. S. Fukuzumi, J. Jung, Y. Yamada, T. Kojima and W. Nam, *Chem. – Asian J.*, 2016, **11**, 1138–1150.
31. J. Lin, Q. Han and Y. Ding, *Chem. Rec.*, 2018, **18**, 1531–1547.
32. S. Fukuzumi, T. Kojima, Y.-M. Lee and W. Nam, *Coord. Chem. Rev.*, 2017, **333**, 44–56.
33. M. Natali, F. Nastasi, F. Puntoriero and A. Sartorel, *Eur. J. Inorg. Chem.*, 2019, 2027–2039.
34. W.-C. Hsu and Y.-H. Wang, *ChemSusChem*, 2022, **15**, e202102378.
35. N. Noll and F. Würthner, *Acc. Chem. Res.*, 2024, **57**, 1538–1549.
36. Y. H. Hong, J. Jung, T. Nakagawa, N. Sharma, Y.-M. Lee, W. Nam and S. Fukuzumi, *J. Am. Chem. Soc.*, 2019, **141**, 6748–6754.
37. K. Ohkubo, A. Fujimoto and S. Fukuzumi, *J. Am. Chem. Soc.*, 2013, **135**, 5368–5371.
38. K. Ohkubo, K. Hirose and S. Fukuzumi, *Chem. – Eur. J.*, 2015, **21**, 2855–2861.
39. H. Kotani, T. Suenobu, Y.-M. Lee, W. Nam and S. Fukuzumi, *J. Am. Chem. Soc.*, 2011, **133**, 3249–3251.
40. S. Hong, Y.-M. Lee, W. Shin, S. Fukuzumi and W. Nam, *J. Am. Chem. Soc.*, 2009, **131**, 13910–13911.
41. Y. H. Hong, Y.-M. Lee, S. Fukuzumi and W. Nam, *Chem*, 2024, **10**, 1755–1765.
42. Y. Hitomi, K. Arakawa and M. Kodera, *Chem. Commun.*, 2014, **50**, 7485–7487.

6 Combination of PSI and PSII Models

6.1 Water Splitting

The stoichiometry of photosynthesis obtained by combination of PSI and PSII is given by eqn (6.1), where H_2O is used as a reductant (an electron and proton source) to reduce $NAD(P)^+$ to yield NAD(P)H regioselectively, accompanied by O_2 evolution under solar irradiation.[1] Because NAD(P)H can reduce water with an $NAD(P)^+$ reduction catalyst to evolve H_2,[2] the stoichiometry of the overall artificial photosynthesis for solar fuel production is given by eqn (6.2), which is water splitting.[1] Water splitting has also been realized using heterogeneous semiconductor photocatalysts.[3–7]

$$2H_2O + NAD(P)^+ \xrightarrow{h\nu} O_2 + 2NAD(P)H + 2H^+ \tag{6.1}$$

$$2H_2O \xrightarrow[\text{PSI + PSII}]{h\nu} 2H_2 + O_2 \tag{6.2}$$

Water splitting using molecular catalysts has been realized by combining a PSI molecular model system and a PSII molecular model system by using two liquid membranes (*vide infra*).[8] Firstly, a PSII model is made with use of *p*-benzoquinone (Q) as a plastoquinone (PQ) analogue, $[(N4Py)Fe^{II}]^{2+}$, as a water oxidation catalyst (WOC) in water/trifluoroethanol (TFE)/toluene in a 3 : 1 : 4 ratio, which are separated in two phases (*e.g.*, water/TFE and toluene), as shown in

RSC Foundations No. 1
Artificial Photosynthesis
By Shunichi Fukuzumi

Published by the Royal Society of Chemistry, www.rsc.org

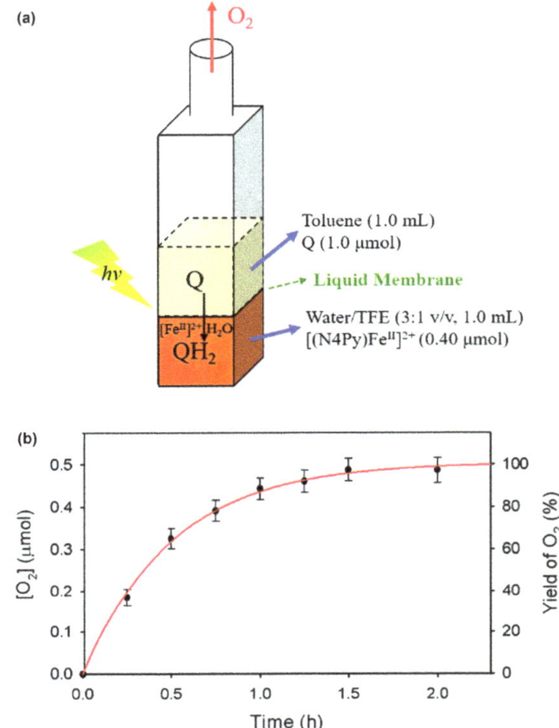

Figure 6.1 (a) A photochemical cell employed for photochemical oxidation of H_2O to O_2 by *p*-benzoquinone (Q: 1.0 μmol) in toluene and water/TFE (3 : 1 v/v) containing $[(N4Py)Fe^{II}]^{2+}$ (0.40 μmol) in water/TFE (3 : 1 v/v). (b) Time profile of O_2 evolution in photochemical oxidation of water to O_2 by Q (1.0 μmol) in toluene and water/TFE (3 : 1 v/v) containing $[(N4Py)Fe^{II}]^{2+}$ (0.40 μmol) under photoirradiation. Reproduced from ref. 8 with permission from American Chemical Society, Copyright 2022.

Figure 6.1a.[8] The same water oxidation catalyst ($[(N4Py)Fe^{II}]^{2+}$) was employed in the case of photochemical oxidation of water to O_2 by DDQ [eqn (5.2)].[8] Q is soluble in toluene, whereas hydroquinone (QH_2) and $[(N4Py)Fe^{II}]^{2+}$ are only soluble in water/TFE (not soluble in toluene).[8] Photoirradiation of a water/TFE/toluene (3 : 1 : 4 v/v/v) solution containing $[(N4Py)Fe^{II}]^{2+}$, which is soluble in water/TFE, and Q, which is soluble in toluene, resulted in evolution of O_2 with 100% yield based on the initial amount of Q according to the stoichiometry of water oxidation by Q [eqn (6.3) and Figure 6.1b], which is the same as that in PSII.[8] The stoichiometric formation of QH_2 (100% yield in water/TFE) in the photochemical water oxidation by Q in water/TFE/toluene solution containing $[(N4Py)Fe^{II}]^{2+}$ and Q was confirmed by the UV-vis absorption spectral changes.[8] Photochemical water oxidation to evolve O_2 occurred similarly when Q was replaced by other *p*-benzoquinone derivatives such as *p*-chloranil (Cl_4Q), 2,3-dichloro-5,6-dicyano-*p*-benzoquinone (DDQ) and 2,5-dimethyl-*p*-benzoqunone (Me_2Q).[8] The stoichiometric formation of the corresponding hydroquinone derivatives (X–QH_2) accompanied by O_2 evolution in TFE/H_2O was also confirmed by the UV-vis absorption spectral measurements.[8] The mechanism of photochemical water oxidation to evolve O_2 by X–Q as a plastoquinone analogue with $[(N4Py)Fe^{II}]^{2+}$ as a water oxidation catalyst may be the same as that reported in acetonitrile (MeCN) containing H_2O (Scheme 5.2).[9] $[(N4Py)Fe^{V}(O)]^{3+}$, which is produced by three step ET oxidation of $[(N4Py)Fe^{II}]^{2+}$ by $^3DDQ^*$, can oxidize water to evolve O_2.[9]

$$2H_2O + 2\,\text{(p-benzoquinone)} \xrightarrow[{[(N4Py)Fe^{II}]^{2+}}]{h\nu} O_2 + 2\,\text{(hydroquinone)} \tag{6.3}$$

A PSI molecular functional model was also constructed using hydroquinone (QH_2) as a plastoquinol analogue, which was produced in a PSII model system, 9-mesityl-10-methylacridinium ion (Acr^+–Mes) as an organic photoredox catalyst (PRC model),[10] and $Co^{III}(dmgH)_2pyCl$ (dmgH = dimethylglyoximate monoanion; py = pyridine)[11] as an H_2 evolution catalyst in water/TFE/toluene (15 : 6 : 19 v/v/v), which are separated in two phases (water/TFE, which is reaction solution after photocatalytic water oxidation and toluene), as shown in Figure 6.2a.[8] Q is soluble in toluene, whereas QH_2, Acr^+–Mes and $Co^{III}(dmgH)_2pyCl$ are soluble in water/TFE. Photoirradiation of water/TFE/toluene (15 : 6 : 19 v/v/v) solution containing QH_2, Acr^+–Mes and $Co^{III}(dmgH)_2pyCl$ resulted in evolution of H_2 with 100% yield based on the initial amount of QH_2 according to the stoichiometry in PSI [Figure 6.2b and eqn (6.4)].[8] The stoichiometric formation of Q in the photocatalytic H_2 evolution in water/TFE/toluene containing QH_2, Acr^+–Mes and $Co^{III}(dmgH)_2pyCl$ [eqn (6.4)] was confirmed by UV-vis absorption spectral measurements.[8] When the stoichiometric formation of Q in the photocatalytic H_2 evolution was completed, Acr^+–Mes remained over 95%, demonstrating the robustness of the photoredox catalyst.[8] Photocatalytic H_2 evolution also occurred similarly when QH_2 was replaced by other hydroquinone derivatives, such as tetrachlorohydroquinone (Cl_4QH_2), 2,3-dichloro-5,6-dicyanohydroquinone ($DDQH_2$) and 2,5-dimethylhydroquinone (Me_2QH_2). Thus,

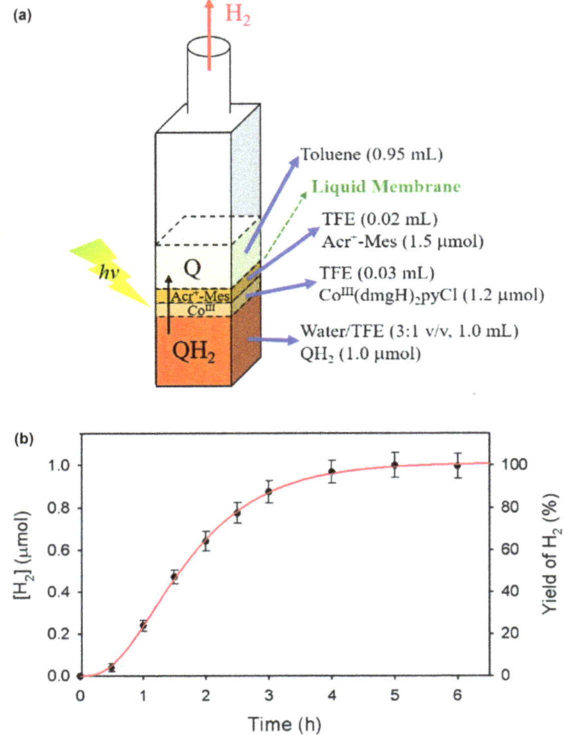

Figure 6.2 (a) A photochemical cell, which was used for photocatalytic H_2 evolution, containing none in toluene (upper) and QH_2 (1.0 μmol), which was produced in the photocatalytic water oxidation, Acr^+–Mes (1.5 μmol) and $Co^{III}(dmgH)_2pyCl$ (1.2 μmol) in water/TFE (3:1 v/v) (lower). (b) Time profile of H_2 evolution in the photocatalytic dehydrogenation of QH_2 with Acr^+–Mes and $Co^{III}(dmgH)_2pyCl$ in water/TFE/toluene under photoirradiation. Reproduced from ref. 8 with permission from American Chemical Society, Copyright 2022.

a series of *p*-benzoquinone/hydroquinone derivatives can be employed as plasotoquinone/plasotoquinol analogues to mimic photocatalytic functions of PSI and PSII.[8]

$$\text{HO–C}_6\text{H}_4\text{–OH} \xrightarrow[\text{Acr}^+\text{–Mes, Co}^{III}(\text{dmgH})_2\text{pyCl}]{h\nu} H_2 + \text{O=C}_6\text{H}_4\text{=O} \quad (6.4)$$

The mechanism of the photocatalytic H_2 evolution, using X–QH_2 as a plastoquinol analogue with Acr^+–Mes as an organic photoredox catalyst and $Co^{III}(dmgH)_2pyCl$ as an H_2 evolution catalyst, may be virtually the same as that previously reported in MeCN containing H_2O (Scheme 4.2).[8,12] Photoexcitation of Acr^+–Mes results in ET from X–QH_2 to $^3(Acr^\bullet–Mes^{\bullet+})$, followed by deprotonation of X–$QH_2^{\bullet+}$ to produce X–$QH^\bullet$ and ET from $Acr^\bullet$–Mes to $Co^{III}(dmgH)_2pyCl$ to afford $[Co^{II}(dmgH)_2pyCl]^-$, accompanied by regeneration of Acr^+–Mes.[12] Then, hydrogen atom transfer from X–$QH^\bullet$ to $[Co^{II}(dmgH)_2pyCl]^-$ occurs to produce X–Q and $[Co^{III}(H)(dmgH)_2pyCl]^-$, which reacts with H^+ to evolve H_2, accompanied by regeneration of $Co^{III}(dmgH)_2pyCl$.[8]

A PSII model (Figure 6.1a) is now combined with a PSI model (Figure 6.2a) to construct a photocatalytic water splitting system, as shown in Figure 6.3a, where the PSI and PSII models are connected in a water/TFE phase, and they are separated by a glass membrane. A glass membrane allows only X–QH_2 to pass, because cationic catalyst molecules cannot pass through the glass membrane composed of silica membrane.[8] Photoirradiation of a water/TFE/toluene (9 : 3 : 5 v/v/v) solution containing Q and $[(N4Py)Fe^{II}]^{2+}$ in the left compartment in Figure 6.3a and Acr^+–Mes and $Co^{III}(dmgH)_2pyCl$ in the right compartment in Figure 6.3a resulted in O_2 evolution from the left compartment (Figure 6.3b) and H_2 evolution from the right compartment (Figure 6.3c) with 100% yield based on the initial amount of Q [eqn (6.5)].[8] The time profile of H_2 evolution showed sigmoidal behaviour, indicating that Q is first reduced by water to QH_2, accompanied by O_2 evolution in the left compartment in Figure 6.3a, and that QH_2 is transferred to the right compartment in Figure 6.3a, where H_2 evolution occurs from QH_2.[8] Thus, there was a time lag for starting H_2 evolution from QH_2.[8] The final yields of O_2 and H_2 were determined to be a 1 : 2 ratio, which agrees with water splitting stoichiometry (Scheme 6.1).[8] The overall water splitting occurred similarly when Q was replaced by other *p*-benzoquinone derivatives (X–Q), such as DDQ, Cl_4Q and Me_2Q.[8]

$$2H_2O \xrightarrow[Acr^+ - Mes, Co^{III}(dmgH)2pyCl]{h\nu/Q/\left[(N4Py)Fe^{II}\right]^{2+}} 2H_2O_2 \tag{6.5}$$

Finally a photocatalytic water splitting system with homogeneous molecular photocatalysts by the X–Q/X–QH_2 cycle was constructed by connecting the two toluene phases, as shown in Figure 6.4a.[8] Photoirradiation of a water/TFE/toluene (3 : 1 : 4 v/v/v) solution containing Q and $[(N4Py)Fe^{II}]^{2+}$ in the left compartment in Figure 6.4a and Acr^+–Mes and $Co^{III}(dmgH)_2pyCl$ in the right compartment in Figure 6.4a resulted in O_2 evolution from the left compartment

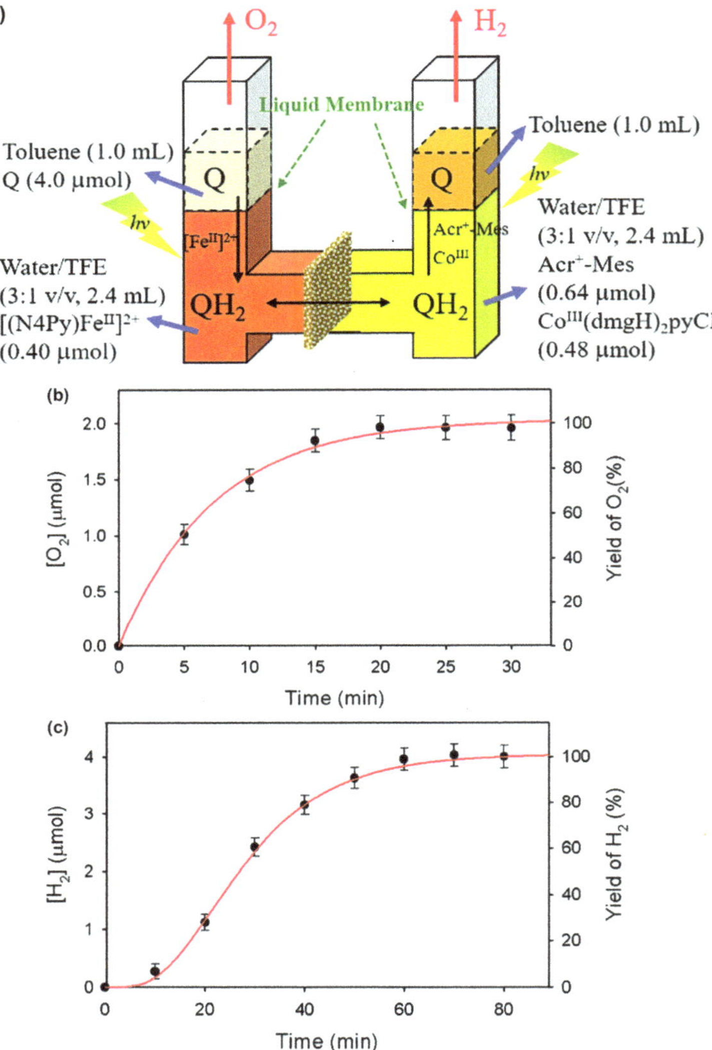

Figure 6.3 (a) A photochemical H-type tube composed of two compartments employed for photocatalytic water splitting: one compartment contains *p*-benzoquinone (Q: 4.0 µmol) in toluene (1.0 mL; upper part) and $[(N4Py)Fe^{II}]^{2+}$ (0.40 µmol) in water/TFE (3 : 1 v/v, 2.4 mL; lower part). The other compartment contains a toluene phase (1.0 mL; upper part) and a water/TFE phase (3 : 1 v/v, 2.4 mL; lower part) containing Acr^+–Mes (0.64 µmol) and $Co^{III}(dmgH)_2pyCl$ (0.48 µmol). The two compartments are connected at the bottom and the water/TFE phase is divided by a glass membrane. (b) Time profile of O_2 evolution in photocatalytic water oxidation (left hand side) and (c) H_2 evolution in the photocatalytic dehydrogenation of QH_2 (right hand side) under photoirradiation. Reproduced from ref. 8 with permission from American Chemical Society, Copyright 2022.

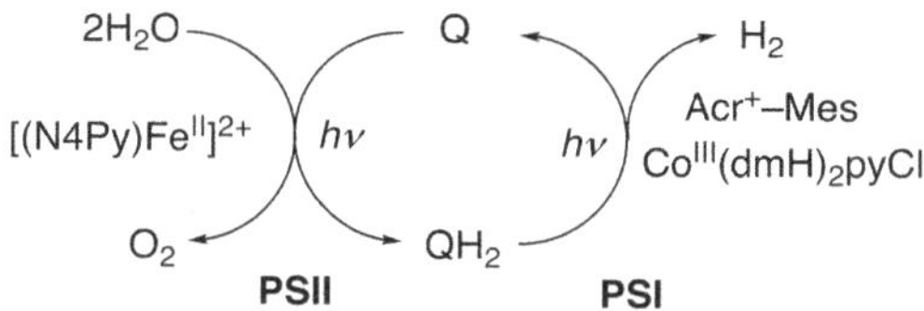

Scheme 6.1 Molecular photocatalytic cycle for water oxidation (O_2 evolution) combined with H_2 by a combination of PSI and PSII models.

(Figure 6.4b) and H_2 evolution from the right compartment (Figure 6.4c).[8] The evolution of O_2 and H_2 under photoirradiation for the initial 30 min was observed in a 1 : 2 ratio, which agrees with the water splitting stoichiometry (Scheme 6.1).[8] At prolonged illumination time, however, the concentration of O_2 decreased gradually, and the formation of H_2 was stopped, because O_2 produced in the left compartment was moved through a glass membrane to the right compartment, where O_2 was reduced by H_2 with an H_2 evolution catalyst to produce hydrogen peroxide.[8] Stoichiometric H_2O_2 production was confirmed by titration with the oxo{[5,10,15,20-tetra-(4-pyridyl)porphyrinato]titanium(IV)} (Ti–TPyP) complex.[13,14] The yield of H_2 was determined to be 1.6 mmol by gas chromatography (GC) and O_2 was partially consumed to produce H_2O_2. The TONs for H_2 evolution of photocatalytic water splitting were determined to be 180 ± 20, 110 ± 15 and 150 ± 20 based on the initial amounts of Q, Acr^+–Mes and $Co^{III}(dmgH)_2pyCl$, respectively.[8] Thus, X–Q acts as an organic photoredox catalyst for a photocatalytic water splitting system, as shown in Scheme 6.1.[8]

Molecular photocatalytic water splitting composed of PSI and PSII models can be further combined with catalytic hydrogenation of CO_2 to achieve CO_2 fixation by H_2O (*vide infra*) as photosynthesis does.

6.2 Reduction of $NAD(P)^+$ to NAD(P)H by H_2O

The same photocatalytic system as water splitting has been employed to achieve the regioselective reduction of $NAD(P)^+$ by H_2O to produce NAD(P)H, combined with water oxidation to evolve O_2, as shown in Figure 6.5.[15] Photoirradiation of a two-phase mixed solution of toluene, TFE and a borate aqueous buffer (0.10 M, pH = 7.0; v/v/v 50 : 1 : 49) of Cl_4Q and $[(N4Py)Fe^{II}]^{2+}$ in the left side cell and also a two-phase mixed solution of toluene, TFE and a borate aqueous buffer (0.10 M, pH = 7.0; v/v/v 50 : 1 : 49) of NAD^+, Acr^+–Mes and

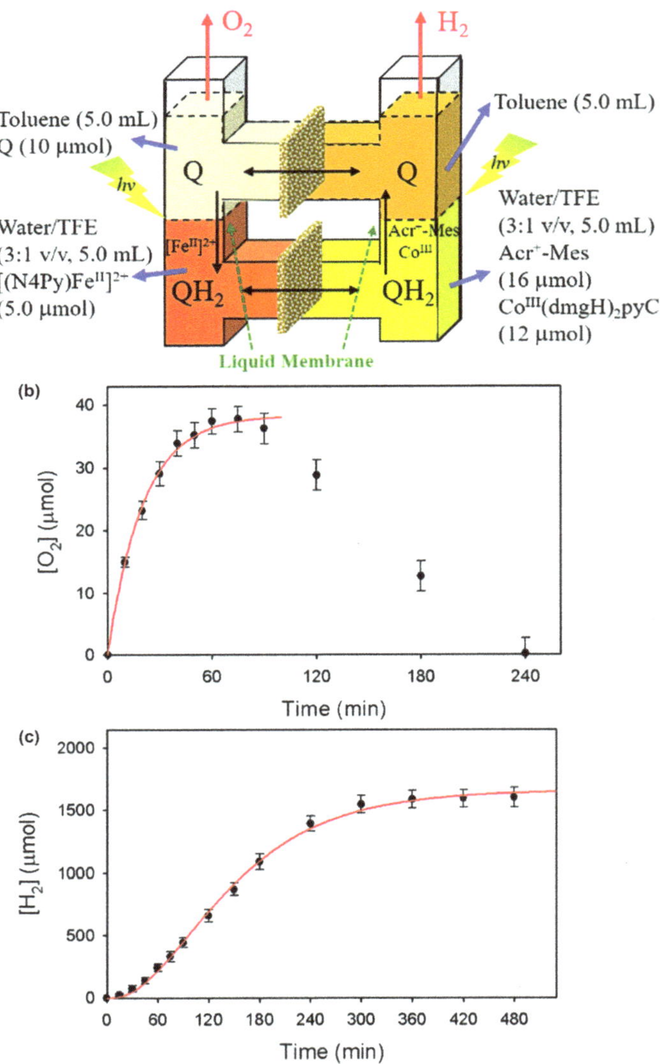

Figure 6.4 (a) A photochemical O-type tube composed of two parts used for photocatalytic water splitting: one part contains *p*-benzoquinone (Q: 10 μmol) in toluene (5.0 mL; upper part) and $[(N4Py)Fe^{II}]^{2+}$ (5.0 μmol) in water/TFE (3:1 v/v, 5.0 mL; lower part). The other part contains a toluene phase (5.0 mL; upper part) and a water/TFE (3:1 v/v, 5.0 mL; lower part) phase containing Acr^{+}–Mes (16 μmol) and $Co^{III}(dmgH)_2pyCl$ (12 μmol). The left toluene phase is connected to the right toluene phase, whereas the left water/TFE phase is connected to the right water/TFE phase and the two connected parts are each divided by a glass membrane. (b) Time course of O_2 evolution in photocatalytic water oxidation (left hand side) and (c) H_2 evolution in the photocatalytic dehydrogenation of QH_2 (right hand side) under photoirradiation. Reproduced from ref. 8 with permission from American Chemical Society, Copyright 2022.

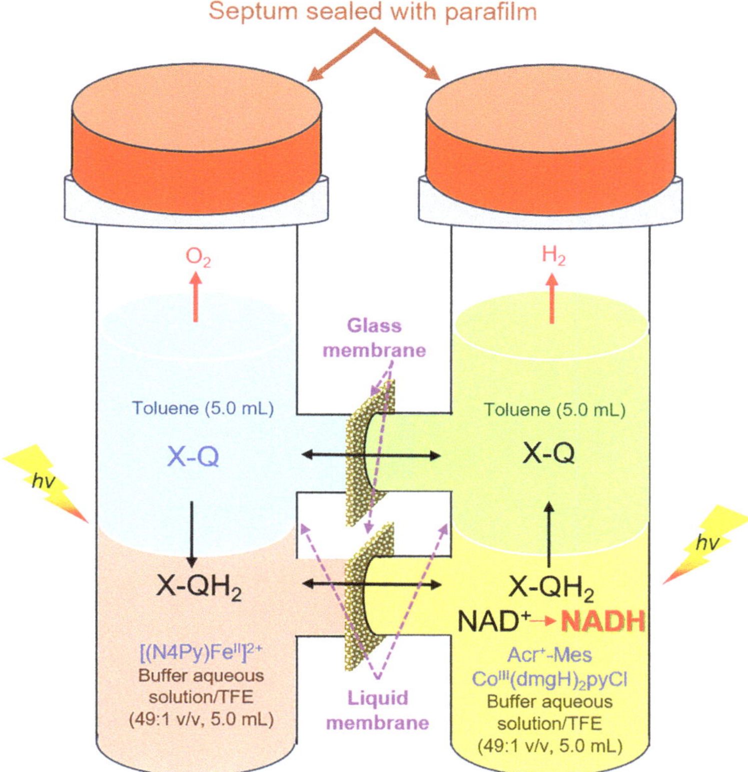

Figure 6.5 A photochemical O-type tube used to produce NADH by photocatalytic regioselective reduction of NAD^+ by H_2O by combining functional models of PSI and PSII using two glass membranes. Reproduced from ref. 15 with permission from American Chemical Society, Copyright 2024.

$Co^{III}(dmgH)_2pyCl$ in the right side cell resulted in regioselective production of NADH (1,4-reduced form) with nearly 100% yield based on the initial amount of NAD^+ together with evolution of H_2 in the right side cell and evolution of O_2 in the left side cell (Figure 6.6a–c).[15] The TON of formation of NADH reached 24 based on Cl_4Q (initial amount), 16 based on $Co^{III}(dmgH)_2pyCl$ and 12 based on Acr^+–Mes. Thus, Cl_4Q, $Co^{III}(dmgH)_2pyCl$ and Acr^+–Mes act as combination catalysts for the overall photocatalytic reduction of NAD^+ by H_2O.[15] The photocatalytic reduction of NADH occurred similarly when Cl_4QH_2 was replaced by hydroquinone (QH_2) and tetramethylhydroquinone (Me_4QH_2). Thus, X–Q plays the same role as plastoquinone that acts as a redox mediator catalyst in the photosynthesis (Scheme 6.2).[15] The amount of NADH produced with H_2 relative to O_2

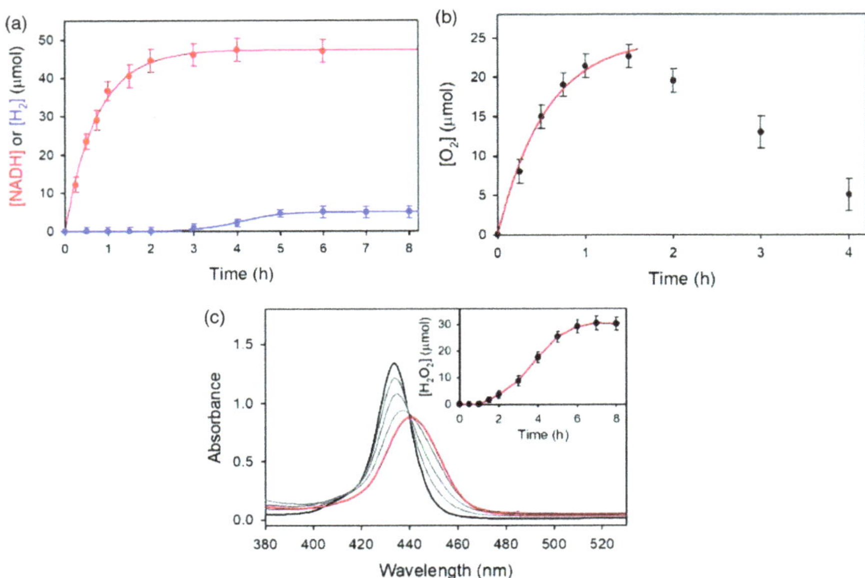

Figure 6.6 (a) Time profiles of formation of NADH and H_2 evolution (right compartment) in the photocatalytic reduction of NAD^+ (200 μmol) and H^+ by Cl_4Q (2.0 μmol), $[(N4Py)Fe^{II}]^{2+}$ (1.0 μmol), Acr^+–Mes (4.0 μmol) and $Co^{III}(dmgH)_2pyCl$ (3.0 μmol) in a toluene/TFE/borate buffer aqueous solution (0.10 M, pH 7.0; 20 mL, 50 : 1 : 49 v/v/v), respectively, accompanied by photocatalytic oxidation of water (left compartment) under photoirradiation. (b) Time profiles of O_2 evolution (left compartment) in the photocatalytic oxidation of water by Cl_4Q (2.0 μmol) with $[(N4Py)Fe^{II}]^{2+}$ (1.0 μmol), Acr^+–Mes (4.0 μmol) and $Co^{III}(dmgH)_2pyCl$ (3.0 μmol) in a toluene/TFE/borate buffer aqueous solution (0.10 M, pH 7.0; 20 mL, 50 : 1 : 49 v/v/v), accompanied by photocatalytic reduction of NAD^+ (right compartment) under photoirradiation. (c) Detection of H_2O_2 produced by the photocatalytic reduction of O_2 by NADH: The titration of H_2O_2 was performed from UV-vis absorption spectra of the mixture solution of Ti-TPyP {oxo[5,10,15,20-tetra-(4-pyridyl)porphyrinato]titanium(IV) complex, 5.0×10^{-6} M} with the 20-fold diluted reaction solution (borate buffer aqueous solution/TFE) in the right compartment of an O-type tube cell at 0 (black line) and 8 h (red line) for production of H_2O_2 in the photocatalytic reduction of NAD^+ (200 μmol) by H_2O with Cl_4Q (2.0 μmol), $[(N4Py)Fe^{II}]^{2+}$ (1.0 μmol), Acr^+–Mes (4.0 μmol) and $Co^{III}(dmgH)_2pyCl$ (3.0 μmol) in a toluene/TFE/borate buffer aqueous solution (0.10 M, pH 7.0; 20 mL, 50 : 1 : 49 v/v/v). Inset shows time profile of formation of H_2O_2 in the photocatalytic reduction of NAD^+ by H_2O under photoirradiation. Reproduced from ref. 15 with permission from American Chemical Society, Copyright 2024.

were obtained in a 2 : 1 ratio in the agreement with the stoichiometry in photosynthesis (Scheme 6.2). At prolonged photoirradiation time, however, the concentration of O_2 decreased gradually and the

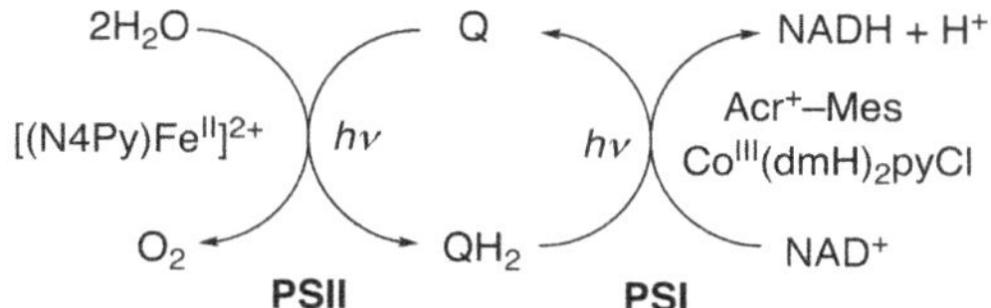

Scheme 6.2 Molecular photocatalytic cycle for production of NADH by photocatalytic regioselective reduction of NAD^+ by H_2O with O_2 evolution achieved by combination of PSI and PSII models.

formation of NADH was stopped, because H_2 evolved by the reaction of NADH with H^+ to evolve H_2 [eqn (6.6)].[15] The reaction solution in the left compartment containing O_2 was passed through a glass membrane to the right compartment, where O_2 was reduced by NADH to produce hydrogen peroxide (Figure 6.6c).[15] When NAD^+ was replaced by its analogues, $NADP^+$ and an NAD^+ model compound, 1-benzyl-3-carbamoylpyridinium cation (BNA^+), NADPH and BNAH were produced as the case of NAD^+ (Figure 6.6a).[15] The photocatalytic regioselective reduction of NAD^+ by H_2O to produce NADH, composed of PSI and PSII models (Scheme 6.2) can be further combined with various NADH dependent enzymatic reactions (*vide infra*).

$$NADH + H^+ \underset{[Co^{III}(dmgH)_2pyCl]}{\rightleftarrows} H_2 + NAD^+ \qquad (6.6)$$

References

1. N. Nelson and C. F. Yocum, *Annu. Rev. Plant Biol.*, 2006, **57**, 521–565.
2. Y. Maenaka, T. Suenobu and S. Fukuzumi, *J. Am. Chem. Soc.*, 2012, **134**, 367–374.
3. Y. Yamaguchi and A. Kudo, *Front. Energy*, 2021, **15**, 568–576.
4. S. Chen, T. Takata and K. Domen, *Nat. Rev. Mater.*, 2017, **2**, 17050.
5. J. D. Xiao, T. Hisatomi and K. Domen, *Acc. Chem. Res.*, 2023, **56**, 878–888.
6. Y. Yan, Z. Chen, X. Cheng and W. Shi, *Catalysts*, 2023, **13**, 967.
7. W.-K. Han, W. Yuan, Z.-G. Gu and Y. Zhao, *ACS Mater. Lett.*, 2024, **6**, 2276–2294.
8. Y. H. Hong, Y.-M. Lee, W. Nam and S. Fukuzumi, *J. Am. Chem. Soc.*, 2022, **144**, 695–700.
9. Y. H. Hong, J. Jung, T. Nakagawa, N. Sharma, Y.-M. Lee, W. Nam and S. Fukuzumi, *J. Am. Chem. Soc.*, 2019, **141**, 6748–6754.
10. S. Fukuzumi, K. Ohkubo and T. Suenobu, *Acc. Chem. Res.*, 2014, **47**, 1455–1464.
11. J. L. Dempsey, B. S. Brunschwig, J. R. Winkler and H. B. Gray, *Acc. Chem. Res.*, 2009, **4**, 1995–2004.
12. Y. H. Hong, Y.-M. Lee, W. Nam and S. Fukuzumi, *Inorg. Chem.*, 2020, **59**, 14838–14846.
13. C. Matsubara, N. Kawamoto and K. Takamura, *Analyst*, 1992, **117**, 1781–1784.
14. K. Takamura, C. Matsubara and T. Matsumoto, *Anal. Sci.*, 2008, **24**, 401–404.
15. Y. H. Hong, M. Nilajakar, Y.-M. Lee, W. Nam and S. Fukuzumi, *J. Am. Chem. Soc.*, 2024, **146**, 5152–5161.

7 Hydrogen Storage

7.1 Hydrogen Peroxide

In contrast to gaseous H_2, hydrogen peroxide (H_2O_2) is a liquid that can be transported over long distances using tankers, trucks and trains. H_2O_2 can also be stored using an H_2O_2 fuel infrastructure, which would be similar to the current gasoline infrastructure. Because a large amount of energy can be released in the disproportionation of H_2O_2 to H_2O and O_2 [eqn (7.1): $\Delta G'^0 = -100\ \text{kJ mol}^{-1}$],[1–3] H_2O_2 is a powerful fuel used in rocket propulsion[4–6] and micromotors.[7–9] In contrast to hydrogen fuel cells that require expensive membranes to separate H_2 and O_2 with expensive Pt catalysts, hydrogen peroxide fuel cells can operate in the absence of an O_2 environment, such as outer space and in water without membranes. The use of hydrogen peroxide instead of O_2 as an oxidant can substantially increase the theoretical voltage of fuel cells and thus improve cell performance.[10–12] Because H_2O_2 is produced from H_2 and O_2 and it is used as a fuel (*vide supra*), H_2O_2 can be regarded as a hydrogen storage compound. It is interesting to note that H_2O_2 is also produced in photosynthesis. Photons absorbed by PSII are used to oxidize water in an oxygen-evolving complex (OEC) connected to PSII. The pH gradient formed over the thylakoid membrane during light-driven electron transport through PSII and PSI is concomitantly utilized by ATP synthase (ATPase) for ATP formation.[13] The reducing power released from PSI *via* PSII by four-electron/four-proton oxidation of H_2O is normally used for the regioselective

RSC Foundations No. 1
Artificial Photosynthesis
By Shunichi Fukuzumi

Published by the Royal Society of Chemistry, www.rsc.org

reduction of $NADP^+$ to NADPH by which CO_2 is reduced in the well-known Calvin–Benson cycle.[14] It is known that PSI can also reduce various electron acceptors including dioxygen (O_2) in the Mehler reaction to produce H_2O_2 *via* disproportionation of a superoxide anion ($O_2^{\bullet-}$) with protons.[15] The electron flow from the photosynthetic electron transport chain to O_2 in the Mehler reaction can eliminate the excess of light energy, thus preventing the electron transport chain from photoinhibition.[16–22] Under normal functional conditions the reduction of O_2 may represent about 5–10% of the total photosynthetic electron flow in C3 plants, but the reduction of O_2 increases to 30% and even higher under stress conditions.[23,24] H_2O_2 is known to regulate plant growth, development, acclimatory and defence responses, being also utilized in metabolic regulation in ways similar to diffusible gases such as NO, CO, or H_2S.[25–28]

$$2H_2O_2 \rightarrow 2H_2O + O_2 \tag{7.1}$$

H_2O_2 has recently attracted much attention as an alternative liquid solar fuel to gaseous H_2.[29–38] H_2O_2 is currently produced in industry by an anthraquinone (AQ) process, in which Pd-catalysed hydrogenation of AQ with H_2 produces anthrahydroquinone (H_2AQ), which is oxidized by O_2 to produce H_2O_2, accompanied by regeneration of AQ.[39,40] The AO process is used to produce most of the world's H_2O_2. However, the high cost of the anthraquinone process resulting from recycling of the extraction solvents, the hydrogenation catalyst and the expensive quinone has precluded the much wider use of H_2O_2 as a fuel and green oxidant. Thus, it is highly desired to produce hydrogen peroxide on site by a one-step process from H_2 and O_2 or by photocatalytic oxidation of H_2O by O_2 using solar energy as performed in natural photosynthesis.

7.1.1 Direct Synthesis of H_2O_2

Extensive efforts have so far been devoted to produce H_2O_2 by direct synthesis of from H_2 and O_2 using heterogeneous catalysts.[41–45] The major problem associated with direct synthesis of H_2O_2 using heterogeneous catalysts comes from the limited selectivity of hydrogen usage due to undesired reactions such as decomposition of H_2O_2 and hydrogenation of H_2O_2.[41–45] In addition, it has been extremely difficult to clarify the heterogeneous catalytic mechanism as compared with the homogeneous catalytic mechanism, in which intermediates can be detected more easily. The first direct synthesis of H_2O_2 from H_2 and O_2 in a homogeneous system in water was reported with use of a water-soluble iridium aqua complex $[Ir^{III}(Cp^*)(4\text{-}(1H\text{-pyrazol-1-yl-}\kappa N^2)\text{benzoic acid-}\kappa C^3)(H_2O)]_2SO_4$ ($[\mathbf{1}]_2SO_4$) and flavin mononucleotide (FMN) in water (*vide infra*).[46]

The carboxylic acid group in **1** is deprotonated to give the carboxylate form, **1-H**$^+$ at pH 6.0 Scheme 7.1).[47] **1-H**$^+$ reacts with H_2 to produce the hydride complex (**2**) (Scheme 7.1).[47] **2** can reduce FMN to $FMNH_2$ at pH 6.0 (Scheme 7.1).[48] $FMNH_2$ is well known to be oxidized by O_2 to regenerate FMN, accompanied by formation of H_2O_2 [eqn (7.2)].[49] The overall catalytic cycle for the direct selective synthesis of H_2O_2 from H_2 and O_2 with **1** and FMN is given by Scheme 7.2.[48] The reduction of H_2O_2 to H_2O by H_2 was also catalysed by **1**.[48] However, the reduction of H_2O_2 was retarded by the presence of Sc^{3+}.[48] Thus, the inhibition of further hydrogenation of H_2O_2 was accomplished by the presence of a strong Lewis acid such as Sc^{3+} because H_2O_2 is known to be stabilized by Sc^{3+}.[50] When **1** (1 μM) and FMN (50 μM) were used for production of H_2O_2 from H_2 and O_2, the turnover number (TON) based on **1** reached 847 after 4 h.[48]

$$FMNH_2 + O_2 \rightarrow FMN + H_2O_2 \tag{7.2}$$

Direct H_2O_2 synthesis was also made possible by using a Rh(II) dimer complex in water.[51] The full catalytic cycle was performed using

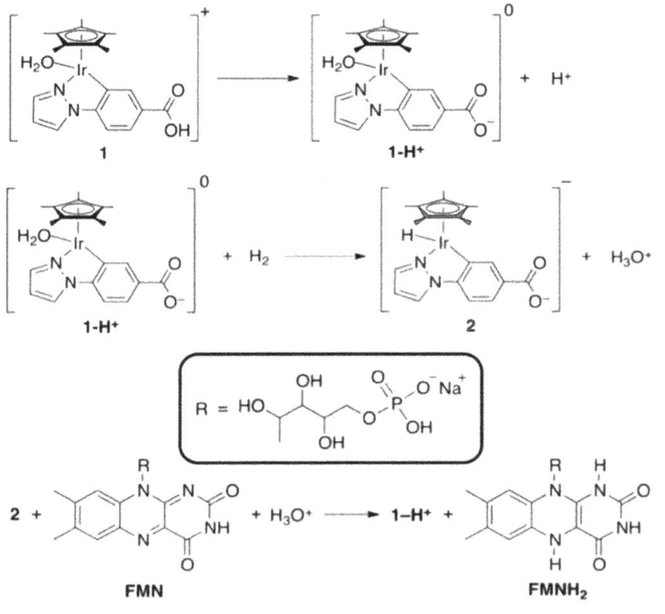

Scheme 7.1

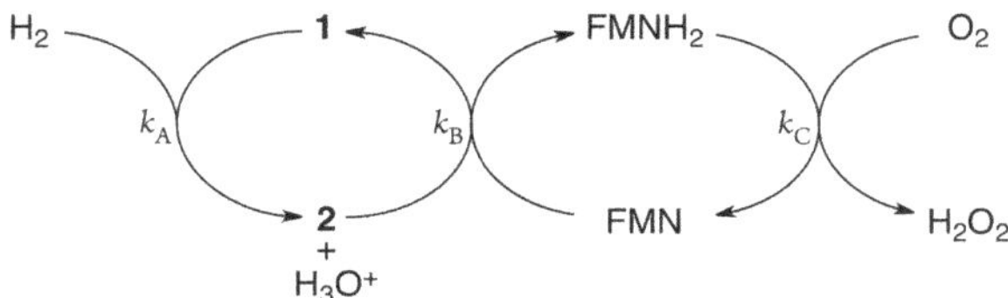

Scheme 7.2 Catalytic cycle for direct synthesis of H_2O_2 from H_2 and O_2 with a water-soluble iridium aqua complex (1) and FMN in water.

an aqueous CH_3COONa solution of a Rh(II) dimer complex (100 μM) under an H_2/O_2 (95/5) atmosphere (0.5–1.9 MPa) at 23 °C for 12 h under dark conditions. The maximum TON reached 910 under an H_2/O_2 (95/5) atmosphere (1.9 MPa) for 12 h at 23 °C.[51] A Rh(II) dimer complex acts as an efficient catalyst for direct synthesis of H_2O_2 from H_2 and O_2 in water.[51] The Rh(II) dimer complex is reduced to two equivalents of the Rh(I) complex by H_2 and the Rh(I) complex reduces O_2 to H_2O_2 in the presence of CH_3COOH.[51]

7.1.2 Photocatalytic Production of H_2O_2 from H_2O and O_2

The first artificial photosynthetic system in which electrons and protons taken from water are used for the two-electron reduction of O_2 to produce H_2O_2 was reported by combining the photocatalytic reduction of O_2 with tris(4,7-dimethyl-1,10-phenanthroline)ruthenium(II) sulphate $[Ru^{II}(Me_2phen)_3](SO_4)$ [eqn (7.3)] and the catalytic oxidation of water by $[Ru^{III}(Me_2phen)_3]^{3+}$ with a water oxidation catalyst (WOC) [eqn (7.4)].[52] The overall reaction is the oxidation of H_2O by O_2 to produce H_2O_2 as observed in photosynthesis, when O_2 is used as an electron acceptor instead of $NADP^+$ [eqn (7.5)].[52]

$$2[Ru^{II}(Me_2phen)_3]^{2+} + O_2 + 2H^+ \xrightarrow{h\nu} 2[Ru^{III}(Me_2phen)_3]^{3+} + H_2O_2 \quad (7.3)$$

$$4[Ru^{III}(Me_2phen)_3]^{3+} + 2H_2O \xrightarrow[WOC]{} 4[Ru^{II}(Me_2phen)_3]^{2+} + O_2 + 4H^+ \quad (7.4)$$

$$2H_2O + O_2 \xrightarrow[Ru^{II},WOC]{h\nu} 2H_2O_2 \quad (7.5)$$

Photoexcitation of $[Ru^{II}(Me_2phen)_3]^{2+}$ resulted in formation of $[Ru^{III}(Me_2phen)_3]^{3+}$ by oxidative quenching of the excited state ($[Ru^{II}(Me_2phen)_3]^{2+}$*: * denotes the excited state) *via* proton-promoted ET to O_2 in the presence of H_2SO_4 to produce $HO_2^•$ which disproportionated to produce H_2O_2.[53] The quantum yield of production of H_2O_2 increased with increasing concentration of H_2SO_4 to reach 72% in the presence of 4.0 M H_2SO_4.[52] When $[Ru^{II}(Me_2phen)_3]^{2+}$ was replaced by $[Ru^{II}(bpy)_3]^{2+}$, the quantum yield became significantly smaller because of the lower reduction potential of $[Ru^{II}(bpy)_3]^{2+}$ ($E^0_{red} = -1.09$ V *vs.* SCE) as compared with that of $[Ru^{II}(Me_2phen)_3]^{2+*}$ ($E^0_{red} = -1.25$ V *vs.* SCE).[52]

With regard to WOC, an efficient but acid-stable catalyst is required for the photocatalytic production of H_2O_2 from H_2O and O_2.[52] $Ir(OH)_3$ nanoparticles were shown to act as an efficient WOC.[52] The photocatalytic production of H_2O_2 from H_2O and O_2 proceeded efficiently using $[Ru^{II}(Me_2phen)_3]^{2+}$ as a photocatalyst for the two-electron reduction of O_2 and $Ir(OH)_3$ as a WOC in an H_2SO_4 aqueous solution (2.0 M) under visible light ($\lambda > 420$ nm) irradiation (Scheme 7.1).[52] Isotope-labelling experiments using $^{18}O^{18}O$ instead of $^{16}O^{16}O$ were conducted to obtain direct evidence for photocatalytic H_2O_2 production, in which the produced H_2O_2 comes from O_2 in the gas phase.[52] The turnover number (TON) based on $[Ru^{II}(Me_2phen)_3]^{2+}$ was determined to be 307 after 18 h photoirradiation.[52] No H_2O_2 production was observed from a reaction solution without $[Ru^{II}(Me_2phen)_3]^{2+}$ or $Ir(OH)_3$ under visible light ($\lambda > 420$ nm).

The effects of metal nitrates $[(M(NO_3)_n)]$ that act as Lewis acids on the photocatalytic production of H_2O_2 from H_2O and O_2 were examined in water instead of H_2SO_4 under otherwise the same reaction conditions.[52] The dependence of the photocatalytic reactivity of H_2O_2 production on the Lewis acidity of metal ions under the same pH conditions (pH 2.8) was examined to show that $Sc(NO_3)_3$ exhibited the highest reactivity in agreement with the strongest Lewis acidity of Sc^{3+}.[52,54,55] The order of the reactivity of metal ions at the same pH conditions (pH 2.8) agrees with a quantitative measure of the Lewis acidity of metal ions.[54,55] The highest reactivity of $Sc(NO_3)_3$ may result from the strongest binding of Sc^{3+} to $O_2^{\bullet -}$ which was produced by ET from $[Ru^{II}(Me_2phen)_3]^{2+*}$ to O_2 to prohibit BET from the $O_2^{\bullet -}$–Sc^{3+} complex[54–56] to $[Ru^{II}(Me_2phen)_3]^{2+}$ as observed in laser-induced transient absorption mesurements.[52] A proposed photocatalytic cycle is shown in Scheme 7.3, where ET from $[Ru^{II}(Me_2phen)_3]^{2+*}$ to O_2 in the presence of Sc^{3+} affords $[Ru^{II}(Me_2phen)_3]^{3+}$ and the $O_2^{\bullet -}$–Sc^{3+} complex, which was detected by EPR.[52] $[Ru^{II}(Me_2phen)_3]^{3+}$ can oxidize water in the presence of a WOC to evolve O_2.[52]

The photocatalytic reactivity of H_2O_2 production from H_2O and O_2 was improved by replacing $Ir(OH)_3$ nanoparticles by a homogeneous WOC, $[Co^{III}(Cp^*)(bpy)(H_2O)]^{2+}$.[52] The TON based on $[Ru^{II}(Me_2phen)_3]^{2+}$ reached 612 after 9 h.[52] The quantum yield of the

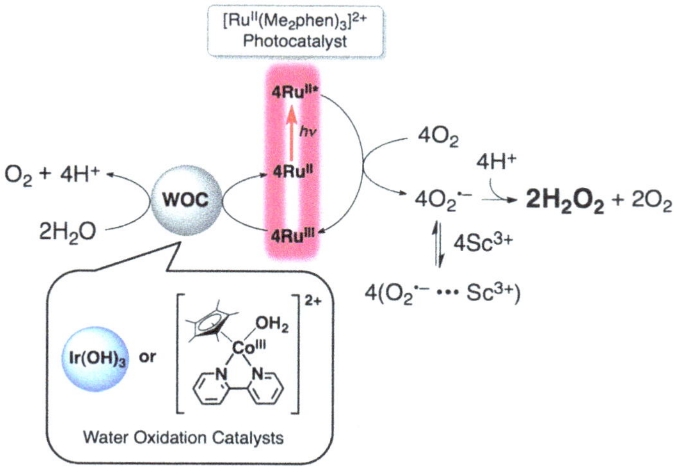

Scheme 7.3 Photocatalytic scheme for H_2O_2 production under photoirradiation of $[Ru^{II}(Me_2phen)_3]^{2+}$ in the presence of $Ir(OH)_3$ or $[Co^{III}(Cp^*)(bpy)(H_2O)]^{2+}$ in H_2O. Reproduced from ref. 52 with permission from the Royal Society of Chemistry.

photocatalytic H_2O_2 production with $[Ru^{II}(Me_2phen)_3]^{2+}$ and $[Co^{III}(Cp^*)(bpy)(H_2O)]^{2+}$ in the presence of $Sc(NO_3)_3$ under photoirradiation at $\lambda = 450$ nm was determined to be 37%, when the energy conversion efficiency of solar energy to chemical energy was obtained to be 0.25%,[52] which is even higher than that of switchgrass (0.2%), a promising crop for biomass fuel.[57]

Nickel ferrite ($NiFe_2O_4$) nanoparticles also act as an efficient WOC in the photocatalytic production of H_2O_2 from H_2O and O_2 with $[Ru^{II}(Me_2phen)_3]^{2+}$ in the presence of Sc^{3+} in water.[58] The size of $NiFe_2O_4$ nanoparticles became smaller during the photocatalytic production of H_2O_2 to exhibit higher catalytic reactivity in the second run as compared with that in the first run.[58] $NiFe_2O_4$ nanoparticles with a diameter 91 nm obtained by the treatment of $NiFe_2O_4$ in an aqueous solution of $Sc(NO_3)_3$ showed 33 times higher catalytic reactivity in the rate of photocatalytic H_2O_2 production as compared with the as-prepared $NiFe_2O_4$ catalyst with a diameter of 1300 nm.[58] It is highly desirable to replace $[Ru^{II}(Me_2phen)_3]^{2+}$ by an earth-abundant metal complex for photocatalytic production of H_2O_2 from H_2O and O_2.

Graphitic carbon nitride (g-C_3N_4) containing electron deficient aromatic diimide units were also reported to act as a photocatalyst for production of H_2O_2 from H_2O and O_2 without any sacrificial electron donor.[59] The g-C_3N_4 catalyst alone exhibited no photocatalytic activity for production of H_2O_2 from H_2O and O_2, because the top of the valence band (VB) lies at approximately 1.4 V (*vs.* NHE at pH 7), which is insufficient for water oxidation.[59] Incorporating the diimide units positively shifts the valence-band potential of the catalysts, while maintaining sufficient conduction-band potential for O_2 reduction.[59] Visible light irradiation of g-C_3N_4 with the diimide units in H_2O with O_2 resulted in production of H_2O_2 by oxidation of H_2O by the photogenerated valence-band holes and selective two-electron/two-proton reduction of O_2 by the conduction band electrons.[59] However, the amount of H_2O_2 produced after 48 h photoirradiation was limited only to 30 μmol.[59] There have been extensive studies on non-metal photocatalysts such as C_3N_4, C dots and resorcinol-formaldehyde resins for H_2O_2 production without any sacrificial reagents.[60–64] These non-metal photocatalysts have so far suffered from problems such as low production rates (about 1 mM h^{-1}) and accumulation amounts (less than 10 mM).[60–64]

Self-assembled tetrakis(4-carboxyphenyl)porphyrin supramolecular nanosheet crystals were also reported to act as a very efficient photocatalyst for production of H_2O_2 from only H_2O and O_2 and with a quantum efficiency of 14.9% at 420 nm and 1.1% at 940 nm.[65] The catalyst achieved a solar-to-chemical conversion efficiency of 1.2% at 328 K when irradiated and heated with simulated sunlight.[65] The concentration of H_2O_2 reached over 30 mM after 24 h.[65] The H_2O_2 production rate decreased when the H_2O_2 accumulation amount was high (*ca.* 20 and 30 mM), because of the H_2O_2 overreaction with e^- or h^+ generated on SA-TCPP photocatalysts, and thermal induced decomposition at high concentrations.[65]

7.1.3 Use of Seawater

Because of the global shortage of clean water, it is highly desirable to utilize earth-abundant seawater instead of pure water as a solvent for production and usage of H_2O_2 as a solar fuel. Efficient photocatalytic production of H_2O_2 from seawater and O_2 in the air was realized by using a two-compartment cell using WO_3 as a photocatalyst for water oxidation and a cobalt chlorin complex (Co^{II}(Ch)) adsorbed on a carbon paper (denoted as CP) as a cathode (Co^{II}(Ch)/CP) for the selective two-electron reduction of O_2, as shown in Figure 7.1.[66] The simulated 1 sun illumination of m-WO_3/FTO in the anode cell afforded the efficient photocatalytic production of H_2O_2 in the cathode cell without any external bias potential.[66] The time courses of photocatalytic H_2O_2 production are shown in Figure 7.2.[66] Virtually

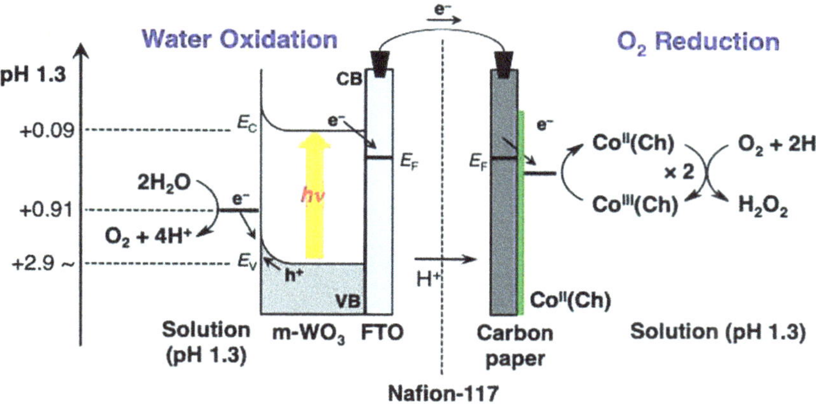

Figure 7.1 Photocatalytic production of H_2O_2 from water and O_2 using an m-WO_3/FTO photoanode and Co^{II}(Ch)/CP cathode in water or seawater under simulated 1 sun (AM 1.5G) illumination. Reproduced from ref. 66, https://doi.org/10.1038/ncomms11470, under the terms of the CC BY 4.0 license, https://creativecommons.org/licenses/by/4.0/.

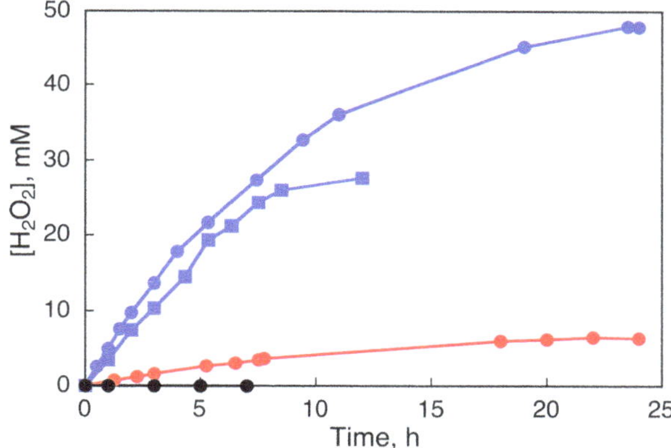

Figure 7.2 Time courses of photocatalytic production of H_2O_2 from H_2O and O_2 with m-WO_3/FTO photoanode and Co^{II}(Ch)/CP cathode in pH 1.3 water (red circle), in pH 1.3 seawater (blue circle) and in an NaCl aqueous solution (pH 1.3) (blue square) under simulated 1 sun (AM 1.5G) illumination. The time course of H_2O_2 production in the absence of Co^{II}(Ch) on carbon paper under simulated 1 sun (AM 1.5G) illumination in pH 1.3 water is shown as black circles. Reproduced from ref. 66, https://doi.org/10.1038/ncomms11470, under the terms of the CC BY 4.0 license, https://creativecommons.org/licenses/by/4.0/.

no H_2O_2 was produced in the absence of Co^{II}(Ch) on the CP electrode, indicating that Co^{II}(Ch) adsorbed on CP efficiently catalyses the two-electron reduction of O_2 to produce H_2O_2,[67] before the charge recombination of photoexcited electron in conduction band and h^+ in VB of WO_3.[66] The rate of photocatalytic production of H_2O_2 in seawater was markedly enhanced compared with that in pure water.[66] After the illumination for 24 h, the amount of produced H_2O_2 in seawater reached *ca.* 48 mM, which was high enough to operate an H_2O_2 fuel cell (*vide infra*).[66]

A similar enhancement of photocatalytic activity was observed in the presence of the same concentration of NaCl in water as that in the seawater.[66] The enhancement effect of Cl^- on the photocatalytic water oxidation can be interpreted by Cl^--promoted water oxidation (*vide infra*).[68] Upon photoexcitation of WO_3, Cl^- is oxidized by a photogenerated hole to produce chlorine (Cl_2) prior to the oxidation of water, as given by eqn (7.9).[68] Cl_2 is converted to HClO depending on the pH of the solution [eqn (7.7)].[68] HClO is decomposed to O_2 and Cl^- under solar irradiation [eqn (7.8)].[68] Thus, the overall water oxidation is catalysed by Cl^- [eqn (7.9)].[68]

$$2Cl^- + 2h^+ \xrightarrow{h\nu} Cl_2 \tag{7.6}$$

$$Cl_2 + H_2O \rightleftarrows HClO + H^+ + Cl^- \tag{7.7}$$

$$2HClO \xrightarrow{h\nu} O_2 + 2H^+ + 2Cl^- \tag{7.8}$$

$$2H_2O + 4h^+ \xrightarrow[Cl^-]{h\nu} O_2 + 4H^+ \tag{7.9}$$

The formation of $HClO/Cl_2$ was confirmed during photocatalytic water oxidation in the presence of Cl^-.[66] The amount of O_2 evolved in seawater in the anode cell after 1 h (12.7 μmol) was more than three times larger than that in water (3.7 μmol).[66] Thus, the enhancement of photocatalytic production of H_2O_2 in seawater results from the Cl^--catalysed photochemical oxidation of water.[66]

The solar energy conversion efficiency for the photocatalytic production of H_2O_2 in seawater was determined to be 0.55% under simulated 1 sun illumination.[66] The best solar energy conversion efficiency was determined to be 0.94% when the illumination intensity was reduced to 0.1 sun.[66] This efficiency is much higher than that of switchgrass (0.2%), which has been considered as a promising crop for biomass fuel.[57]

When surface modified $BiVO_4$ with iron(III) oxide(hydroxide) (FeO(OH)) and Co^{II}(Ch) were employed as a WOC in the photoanode and as an O_2 reduction catalyst in the cathode, respectively, the highest solar energy conversion efficiency was achieved to be

6.6% under simulated solar illumination adjusted to 0.05 sun after 1 h of photoirradiation.[69]

7.1.4 Hydrogen Peroxide Fuel Cells

When an H_2O_2 fuel cell uses H_2O_2 as a fuel and O_2 as an oxidant, the maximum output potential theoretically achievable is 0.55 V, which is less than half of those of an H_2 fuel cell (1.23 V) and a direct methanol fuel cell (1.21 V).[70,71] When H_2O_2 is used as both an oxidant and a reductant, however, the theoretical output potential becomes as high as 1.09 V, which is comparable to those of H_2 and methanol fuel cells.[70,71] The reactions occurring at an anode and a cathode of the H_2O_2 fuel cell are given by eqn (7.10)–(7.12).[70,71]

$$\text{Anode: } H_2O_2 \rightarrow O_2 + 2H^+ + 2e^- \quad 0.68\text{ V } vs.\text{ NHE} \tag{7.10}$$

$$\text{Cathode: } H_2O_2 + 2H^+ + 2e^- \rightarrow 2H_2O \quad 1.77\text{ V } vs.\text{ NHE} \tag{7.11}$$

$$\text{Total: } 2H_2O_2 \rightarrow O_2 + 2H_2O \quad 1.09\text{ V } vs.\text{ NHE} \tag{7.12}$$

H_2O_2 fuel cells with a simple one-compartment structure without membrane can be constructed by employing an anode and a cathode for H_2O_2 oxidation and reduction, respectively (Figure 7.3).[70,71] Such a one-compartment structure without membrane is certainly more promising for development of low-cost fuel cells than a two-compartment structure with membranes.[70,71]

The first one-compartment H_2O_2 fuel cell was reported by using an Au plate as an anode and an Ag plate as a cathode.[72] The H_2O_2 fuel cell was operated in an alkaline solution, in which the output potential was only about 0.1 V, which is much lower than the theoretical voltage of 1.09 V.[72,73] Use of an acidic solution is preferred because H_2O_2 is efficiently produced *via* the two-electron/two-proton reduction of O_2 with under acidic conditions.[74] The one-compartment H_2O_2 fuel cell operated in an acidic solution (pH 3) was reported by using an iron phthalocyanine complex as a cathode and a Ni mesh as an anode.[74] Although the power density was only 10 μW cm^{-2}, the open circuit potential reached to 0.55 V.[74] A higher open circuit potential (0.78 V) and a high power density (1.2 mW cm^{-2}) were obtained for an H_2O_2 fuel cell by using $Fe^{II}{}_3[Co^{III}(CN)_6]_2$ on a carbon cloth as the cathode and a Ni mesh as the anode under acidic conditions at pH 1.[75] The power density of a one-compartment H_2O_2 fuel cell employing a Ni mesh and $[Fe^{II}(H_2O)_2]_3[Co^{III}(CN)_6]_2$ as an anode and a cathode,

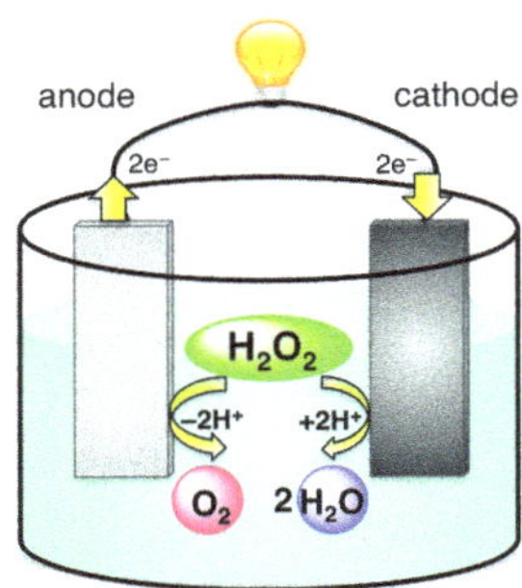

Figure 7.3 A schematic drawing of a one-compartment H_2O_2 fuel cell.

respectively, was dramatically improved to 9.9 mW cm^{-2} by the addition of Sc^{3+} ions to an aqueous H_2O_2 fuel.[76]

The chemical energy of H_2O_2 produced by the photocatalytic oxidation of seawater by O_2 in air was converted to electrical energy through a H_2O_2 fuel cell composed of $Fe^{II}{}_3[Co^{III}(CN)_6]^{2-}$ modified carbon cloth cathode and a nickel mesh anode in a one-compartment cell.[66] The reaction solution (seawater, pH 1.3) containing *ca.* 48 mM of H_2O_2 in the cathode was transferred to the H_2O_2 fuel cell. The cell exhibited an open-circuit potential and a maximum power density of 0.78 V and 1.6 mW cm^{-2}, respectively (Figure 7.4).[66] The energy conversion efficiency of the H_2O_2 fuel cell was determined to be *ca.* 50% by the measurement of output energy as electrical energy *vs.* consumed chemical energy H_2O_2, which is comparable to the efficiency of a H_2 fuel cell.[66] As described above, the proof of concept has been demonstrated for the production of H_2O_2 from seawater and O_2 in the air using solar energy and its direct use on site as a fuel in one-compartment H_2O_2 fuel cells without membranes.[66]

7.2 Formic Acid

Formic acid (HCOOH), which is produced by the two-electron/two-proton reduction of CO_2 has received much attention, because HCOOH is a liquid that is easy to store and carry as compared with gaseous H_2.[77–83] In addition, H_2 can be easily regenerated from HCOOH whenever it is needed on site.[77–83] The selective two-electron reduction of CO_2 to HCOOH (not to CO) has been made possible by selecting metal complex catalysts.[77–83] For example, the hydrogenation of CO_2 to HCOOH in aqueous solutions was found to be catalysed by metal–aqua complexes such as a Ru–aqua complex,

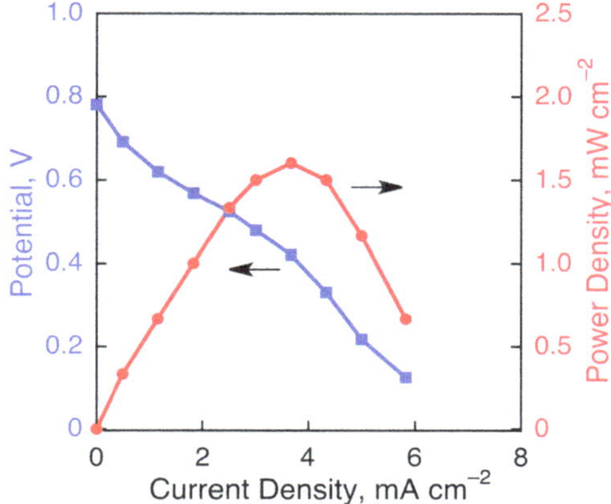

Figure 7.4 *I–V* (blue) and *I–P* (red) curves of the one-compartment H_2O_2 fuel cell with a Ni mesh anode and $Fe^{II}_3[Co^{III}(CN)_6]_2$/carbon cloth cathode in the reaction solution containing H_2O_2 (47.9 mM) produced by photocatalytic reaction in seawater as shown in Figure 7.2 (blue circles). Reproduced from ref. 66, https://doi.org/10.1038/ncomms11470, under the terms of the CC BY 4.0 license, https://creativecommons.org/licenses/by/4.0/.

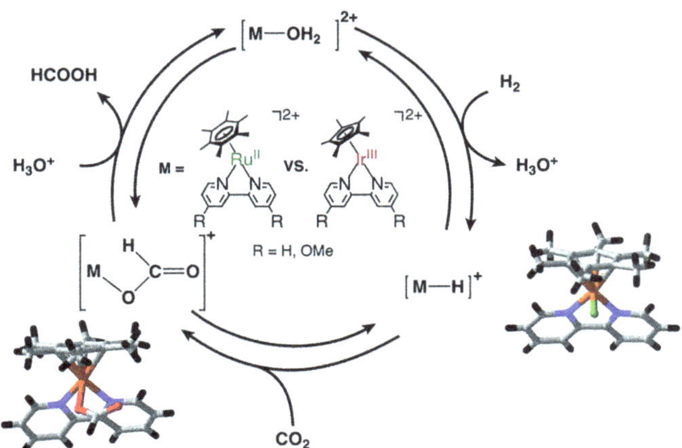

Scheme 7.4 Catalytic mechanism of CO_2 reduction to HCOOH by H_2 with a metal–aqua complex catalyst in water.

$[(\eta^6\text{-}C_6Me_6)Ru^{II}(bpy)(OH_2)]^{2+}$ (bpy = 2,2′-bipyridine), and an Ir–aqua complex, $Ir[(\eta^5\text{-}C_5Me_5)Ir^{III}(bpy)(OH_2)]^{2+}$.[84–86] The catalytic mechanism is shown in Scheme 7.4 where a metal–aqua complex $[(M\text{-}OH_2]^{2+}$ reacts with H_2 to produce a metal–hydride complex ($[M^{III}\text{-}(H^-)]^+$) with release of H_3O^+.[84–86] $[M^{III}\text{-}(H^-)]^+$ reacts with CO_2 to afford the formate complex ($[M^{III}\text{-}OC(H)O]^+$), which reacts with H_3O^+ to yield HCOOH, accompanied by regeneration of

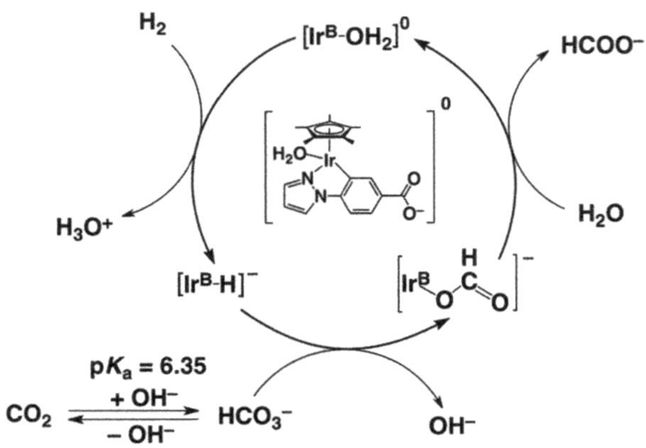

Scheme 7.5 Catalytic mechanism of the hydrogenation of CO_2 to HCOOH with [Ir^B–OH_2].

$[M–OH_2]^{2+}$.[84–86] The catalytic interconversion between H_2 and HCOOH with use of $[M–OH_2]^{2+}$ is reversible, as shown in Scheme 7.4, where the X-ray crystal structures of $[M–H]^+$ (M = Ir, Ru) and $[M–OC(H)O]^+$ are shown.[84–86] In this case, a high pressure (P_{H_2}/P_{CO_2} = 5.5/2.5 MPa) was required for catalytic CO_2 hydrogenation at 313 K.[85]

A water-soluble Ir(III)–aqua complex ($[Ir^AOH_2]^+$: $[(Cp^*)Ir^{III}(4\text{-}(1H\text{-}$pyrazol-1-yl-κ$N^2$)benzoic acid-κ$C_3$)($H_2O$)$]_2SO_4$) was reported to catalyse the hydrogenation of CO_2 to HCOOH under atmospheric pressure of H_2 and CO_2 at room temperature for the first time.[87,88] Bicarbonate (HCO_3^-) [not carbonate (CO_3^{2-}) or CO_2] is reduced by H_2 with use of a deprotonated Ir(III)–aqua complex ([Ir^B–OH_2]) to yield formate at pH 8.8 (Scheme 7.5).[88] At low pH (acidic conditions), however, the catalytic cycle was reversed and dehydrogenation of HCOOH occurred with use of the same Ir(III)–aqua complex to yield CO_2 and H_2 in a 1 : 1 molar ratio (Scheme 7.6).[88] The Ir–hydride complex is the key intermediate for the interconversion between H_2 and HCOOH, being detected by ^{1}H NMR.[88] Extensive studies on the catalytic hydrogenation of CO_2 to HCOOH and dehydrogenation of HCOOH have so far been performed to improve the catalytic activity.[89–98]

7.3 Carbon Monoxide

A reverse water–gas shift reaction (RWGSR), which is the hydrogenation of CO_2 to carbon monoxide (CO), was reported to be catalysed by a series of mononuclear Ru halogen carbonyl complexes, [PPN][$RuX_3(CO)_3$] (PPN = bis(triphenylphosphine)iminium, X = Cl, Br,

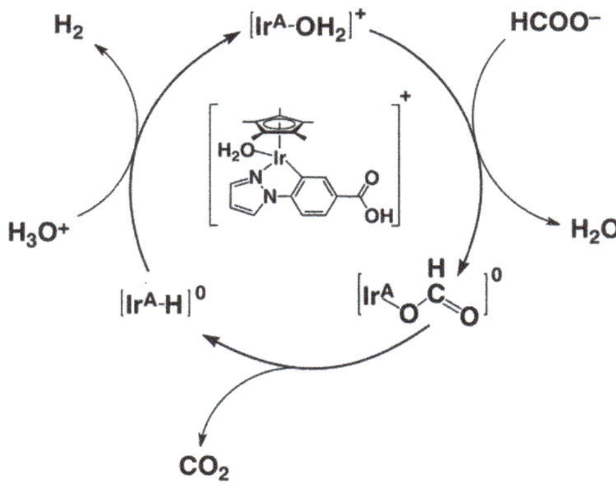

Scheme 7.6 Catalytic mechanism of H_2 evolution by dehydrogenation of HCOOH with $[Ir^B{-}OH_2]$.

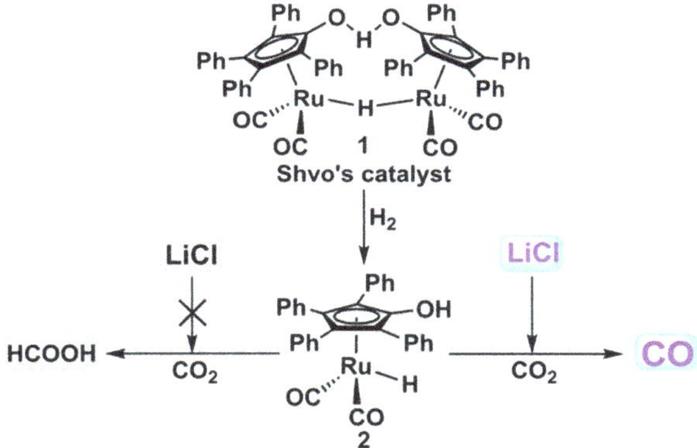

Scheme 7.7 Hydrogenation of CO_2 to CO by Shvo's catalyst with LiCl. Reproduced from ref. 100 with permission from American Chemical Society, Copyright 2021.

and I).[99] When $[PPN][RuCl_3(CO)_3]$ was used as a catalyst in the presence of [PPN]Cl in NMP, the reaction of H_2 (6 MPa) and CO_2 (2 MPa) at 160 °C for 5 h formed CO in a yield of 87 TON based on a Ru atom. First, $[RuCl_3(CO)_3]^-$ releases CO by ligand exchange with the solvent to form $[RuCl_3(CO)_2(solvent)]^-$.[99] The reaction of $[RuCl_3(CO)_2(solvent)]^-$ with hydrogen forms a Ru–hydride complex, $[RuCl_2(CO)_2H(solvent)]^-$, with the release of HCl.[99] After coordination of CO_2, the coordinated CO_2 reacts with a hydrogen ligand and a proton formed in step 2 to give a carbonyl ligand and water.[99] Finally, ligand exchange of $RuCl_2(CO)_3$ with Cl^- regenerates the starting complex.[99]

Shvo's catalyst, shown in Scheme 7.7, can hydrogenate CO_2 to HCOOH selectively with no generation of CO.[100] However, addition of LiCl resulted in switch of the selectivity of the product from HCOOH toward CO even at extremely low temperatures (45–60 °C).[100] Under the optimized conditions, the TON_{CO} reached 1555 at 160 °C, which is much higher than that of state-of-the-art Ru-based homogeneous catalytic systems.[100] The catalytic mechanism was proposed as shown in Scheme 7.8, where the splitting of Shvo's catalyst by H_2 affords a Ru–H complex **2** in the presence of LiCl, followed by the substitution of the carbonyl ligand with LiCl, which affords a Li^+-substituted Ru–H complex S3.[100] The coordination with Cl^- instead of CO enhances the nucleophilicity of the Ru–H complex, which changes the manner of CO_2 insertion into the

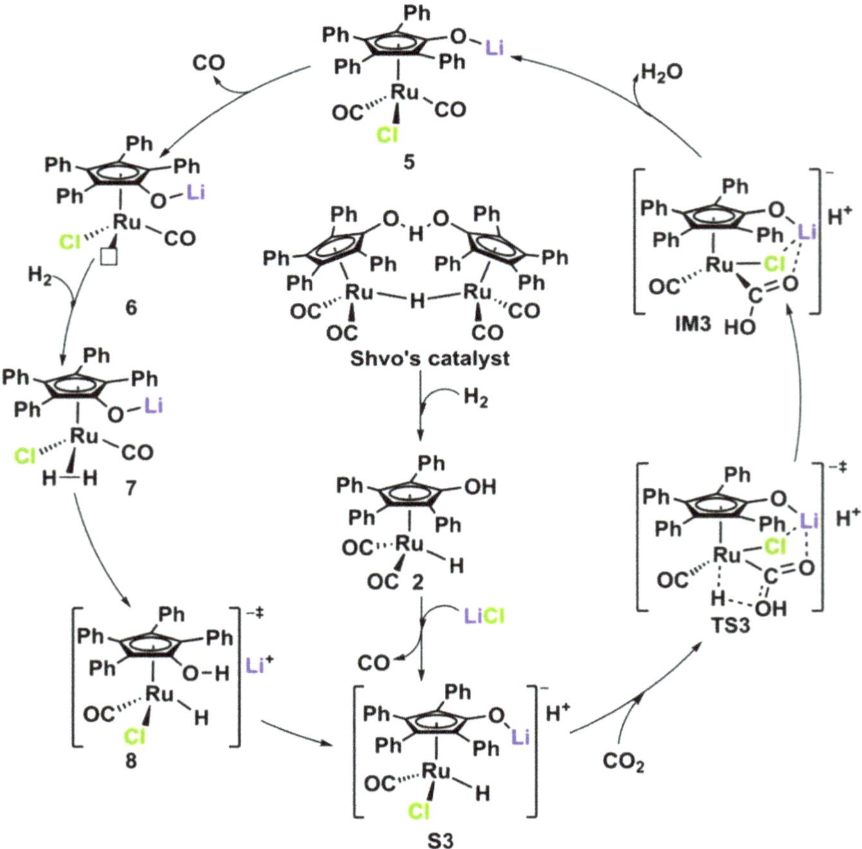

Scheme 7.8 Plausible mechanism for LiCl-assisted hydrogenation of CO_2 to CO. Reproduced from ref. 100 with permission from American Chemical Society, Copyright 2021.

Ru–H bond, affording a Ru(COOH) intermediate (IM3) rather than a Ru(OCOH) complex.[100]

This is followed by the dehydration of IM3 to afford an intermediate **5**, which releases CO to give the undercoordinated intermediate **6**.[100] The heterolytic splitting of H_2 on the intermediate **6** results in regeneration of S3 to complete the catalytic cycle.[100] The intermediate **5** was air-stable enough to be prepared by the reaction of either Shvo's catalyst or complex **2** with LiCl under a N_2 atmosphere.[100] Intermediate **5** in the NMP solvated dimeric form was successfully characterized by X-ray crystallography.[100] Under the optimal conditions for the CO_2 hydrogenation to CO with a higher loading of Shvo's catalyst, the intermediate **5** was detected and isolated.[100] At the same time, the other molecular ion peak with an *m*/*z* of 540.0210 was detected, being assigned to the anionic moiety of the intermediate **6**, which was produced *via* one carbonyl ligand dissociating from the intermediate **5**.[100] The air-unstable active Ru—H species **8** or S3 was also detected by *ex situ* ^{1}H NMR.[100] These results supported the proposed mechanism in Scheme 7.8.[100]

The role of ionic liquids (ILs) and the interaction pattern between Shvo's catalyst and ILs have been revealed by ^{1}H NMR and crystallography measurements of the catalytic hydrogenation of CO_2.[101] ILs promote the dissociation of Shvo's catalyst to enhance the rate of production of CO.[101] The CO that is produced is subsequently used in the tandem hydroformylation-reduction of alkenes to produce valuable alcohols.[101] In the absence of ILs, however, HCOOH was selectively produced, being used for production of formamides by *N*-formylation of most primary or secondary amines.[101]

7.4 Formaldehyde

Formaldehyde (HCHO) is an important ingredient and key building block in many industries such as resins, polymers, cosmetics, adhesives and paints.[102–104] HCHO is industrially produced *via* three stages (the formox process): (a) steam reforming of natural gas to syngas, (b) methanol (CH_3OH) synthesis and (c) partial oxidation of CH_3OH to HCHO.[105] However, large units for combustion, compression and purification and all high temperature reactions in the formox process are highly energy-intensive, resulting in high economic costs and ecological deficiencies.[105] A more environmentally friendly and efficient route for HCHO production is highly desired for a sustainable society.[104] Hydrogenation of CO_2 may be a promising way to produce

HCHO under mild conditions.[104] However, the hydrogenation of CO_2 to HCHO by H_2 in gas-phase reactions is thermodynamically unfavourable, because it is a strongly endothermic reaction at ambient conditions ($CO_2(g) + 2H_2(g) \rightarrow HCHO(g) + H_2O(g)$: $\Delta H^0 = 39.8\ kJ\ mol^{-1}$).[106] Tanksale and co-workers first reported the conversion of syngas ($CO + H_2$) to HCHO in the aqueous phase.[107,108] Although gas-phase hydrogenation of CO by H_2 for HCHO production is also thermodynamically unfavoured to afford low yields of HCHO (<0.2%), the hydrogenation of CO_2 in the liquid rather than the gaseous phase caused the reaction to become thermodynamically feasible to yield 19.14% of CO conversion and 100% of a selectivity toward HCHO using a Ru–Ni/Al_2O_3 catalyst in water at 80 °C and 100 bar.[107]

A much more active heterogeneous catalyst, a layered double hydroxide (LDH)-supported ruthenium (Ru) catalyst was recently reported for HCHO production *via* liquid-phase hydrogenation of CO_2.[106] The highest HCHO yield of 583 $mmol\ L^{-1}\ g_{cat}^{-1}$ and high selectivity toward HCHO were achieved at 30 °C with 10 bar of CO_2 and 10 bar of H_2 in water.[106] One important reason for the high catalytic activity of Ru/LDH-red is ascribed to an electron-rich Ru.[106] The proposed catalytic mechanism based on DFT calculations suggests that *HCOOH species is formed from CO_2 and H_2 on Ru/LDH-red.[106] Then, the dissociative adsorption of H_2 occurs on Ru cluster, inducing the formation of *HCO intermediate and release of one H_2O molecule.[106] The adsorbed *HCO species continue to react with another *H until HCHO is formed on the catalyst surface.[106] The step from *HCOOH species towards *HCO on the catalyst surface is suggested to be the rate limiting step (RLS) during the whole CO_2 reduction process.[106] This hybrid electronic state of Ru and a deep affinity with CO_2 of LDH in the Ru/LDH-red catalyst contribute to overcoming the energy barrier for the formation of *HCO species accompanied with H_2 dissociation, affording its superior catalytic performances even at ambient conditions.[106]

HCHO as well as HCOOH can be regarded as an organic hydrogen carrier, and water-soluble ruthenium complexes were reported to catalyse selective hydrogen production from aqueous formaldehyde under mild conditions.[109] Among the water soluble Ru complexes in Figure 7.5, the biphenyldiamine ruthenium complex **4** afforded the highest yield of H_2 (42%) at 50 °C.[109] The H_2 yield increased to 95% in 7 h at 95 °C.[109] Complex **6**, with four methyl substituents, afforded a much lower yield (3% H_2), compared to the diamine catalyst 4.[109] This suggests the importance of the N–H group in the dehydrogenation of aqueous formaldehyde.[109] The catalyst 4 afforded a high TOF of 8300 h^{-1} with a TON of 24 000 after 100 h at 95 °C.[109]

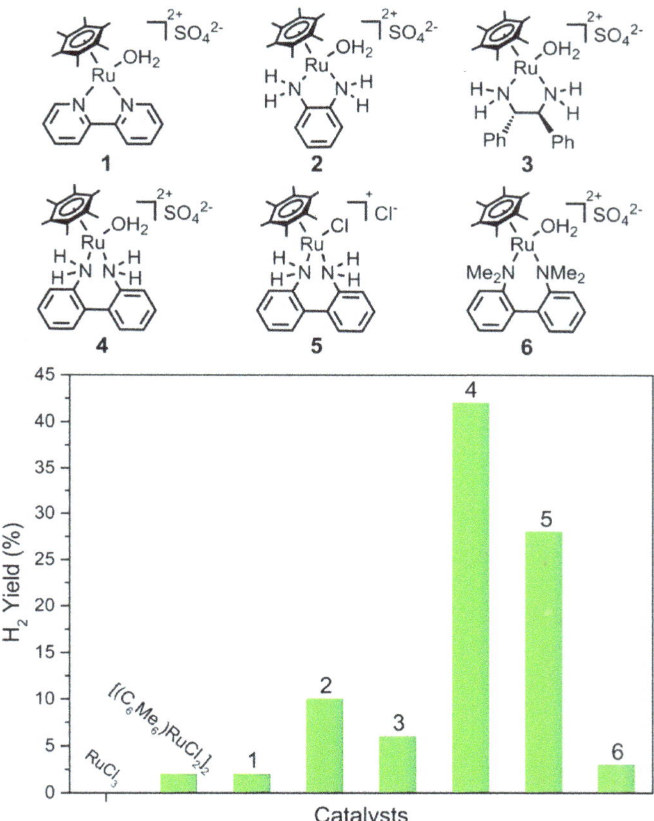

Figure 7.5 Screening of water-soluble Ru catalysts for H_2 production from formaldehyde and water. Conditions: formaldehyde (0.5 mmol), Ru catalyst (10 μmol), H_2O (5 mL), 50 °C, 7 h, $[(C_6Me_6)RuCl_2]_2$ (5 μmol). Reproduced from ref. 109 with permission from American Chemical Society, Copyright 2018.

An arene Ru(II) complex, $[(\eta^6\text{-}p\text{-cymene})RuCl(L)]^+Cl^-$ (L = 4,4′-((2-methoxyphenyl-methylene)bis(2-ethyl-5-methyl-1*H*-imidazole), was also reported to catalyse H_2 evolution from HCHO in water with an exceptionally high turnover number (TON) of >20 000 at 90 °C.[110] It is known that formaldehyde is primarily present as stable methanediol in aqueous solution. The Ru–aqua species [Ru]–A is formed by the loss of a chloro ligand from the [Ru] catalyst and the subsequent deprotonation of the ligand.[110] Then, the Ru–methanediol species ([Ru]–B) is formed by the reaction of the Ru–aqua species [Ru]–A with methanediol, followed by the release of H_2 to produce the Ru–formato species ([Ru]–C).[110] The [Ru]–C complex undergoes decarboxylation to form the Ru–hydride species ([Ru]–D).[110] Finally, the Ru–aqua species

[Ru]–A was regenerated from [Ru]–D *via* a proton-assisted hydrogen release step to complete the catalytic cycle.[110]

Paraformaldehyde, which is the polymerization product of formaldehyde (*i.e.*, $HO(CH_2O)_nH$, where $n = 8–100$), is regarded as a convenient solid H_2 carrier with a higher energy density (6.7%) than formaldehyde (4.4%).[111] A water-soluble Ir(III)–aqua complex was reported to catalyse H_2 evolution from paraformaldehyde in water at 298 K.[111] At pH 11, the Ir–aqua complex is converted to the Ir–hydroxo complex ($[Ir–OH]^-$), which acts as an efficient H_2 evolution catalyst from paraformaldehyde. Paraformaldehyde reacts with $[Ir–OH]^-$ to produce $HO(CH_2O)_{n-1}H$ and the methanediol adduct ($[Ir–OCH_2OH]^-$), which undergoes β-hydrogen elimination to afford the hydride complex ($[Ir–H]^-$) and formic acid.[111] The formation of the methanediol adduct and the hydride complex has been confirmed by ^{1}H-NMR and ESI-MS measurements.[111] The reaction of $[Ir–H]^-$ with H_2O results in H_2 evolution, accompanied by regeneration of $[Ir–OH]^-$.[111] $[Ir–OH]^-$ also reacts with formate to produce $[Ir–H]^-$ and CO_2 *via* β-hydrogen elimination.[111] H_2 is also evolved by the reaction of $[Ir–H]^-$ with H_2O, accompanied by regeneration of $[Ir–OH]^-$.[111] Homogeneous and heterogeneous catalysts have so far been developed for H_2 evolution from formaldehyde in water.[112–115]

7.5 Methanol

Methanol is currently produced in industry from syngas, being in high demand as a promising alternative fuel and useful bulk chemical. Extensive efforts have so far been devoted to develop heterogeneous catalysts for the thermocatalytic hydrogenation of CO_2 to methanol.[116–119] However, these catalysts require harsh reaction conditions due to the high thermodynamic stability and low reactivity of CO_2.[116–119] Thus, homogeneous catalysts, which operate at lower temperatures, in the hydrogenation of CO_2 to methanol have attracted increasing attention.[120–125]

Himeda and coworkers have reported production of methanol *via* the hydrogenation of CO_2 under mild reaction conditions (30 °C, 5 MPa (TON 2.0) and 0.5 MPa, 70 °C (TON 3.0)) using multinuclear Ir complexes (Figure 7.6) under gas–solid phase reaction conditions.[126] Mononuclear catalyst **1** (9 μmol) in water without base generated formic acid, but the concentration of formic acid was immediately stopped.[126] When the reaction was performed with use of **2**-*m* as an amorphous powdery solid, the yield of methanol increased with the reaction time.[126] Methanol was exclusively produced without

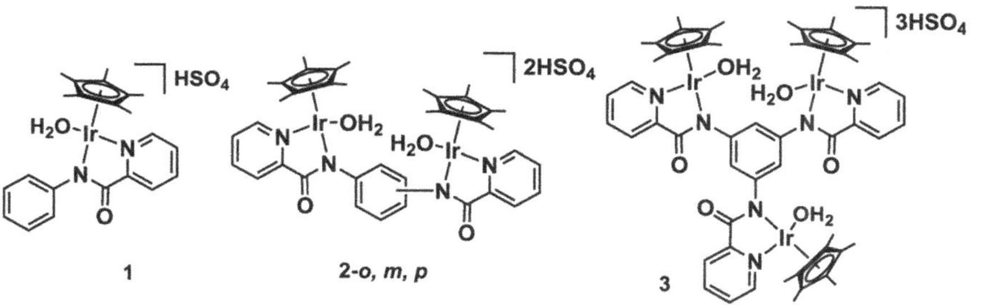

Figure 7.6 Mononuclear (1), dinuclear (2) and trinuclear (3) Ir catalysts employed for the hydrogenation of CO_2 to methanol under mild conditions. Reproduced from ref. 126 with permission from American Chemical Society, Copyright 2021.

detection of CO and CH_4 in the gas phase.[126] On the other hand, a negligible amount of formic acid was obtained in the residual catalyst, and no formaldehyde (methanediol) was detected.[126] The *ortho*-substituted catalyst (**2**-*o*) showed similar activity to that of **2**-*m*, whereas the *para*-substituted catalyst **2**-*p* showed no catalytic reactivity.[126] The trinuclear catalyst **3** afforded a similar amount of methanol to that of **2**-*m*. Such quite different activities observed among the mono and multinuclear Ir catalysts may result from the relative configuration of each iridium species to allow multiple hydride transfer to generate formaldehyde or methanol.[126] A recycling experiment of **2**-*m* using five cycles afforded 0.507 mmol of methanol with a TON of 113 under 4 MPa of H_2/CO_2 (3 : 1) at 60 °C and 336 h for each run.[126]

Homogeneous ruthenium and rhodium RAPTA-type complexes [Ru(η^6-*p*-cymene)X_2(PTA)] [X = I (**1**), Cl (**2**)] and [Rh(η^5-C_5Me_5)X_2(PTA/PTA-BH_3)] (X = Cl (**3**), H (**4**) and PTA-BH_3, H (**5**)) represent the first example of CO_2 hydrogenation to methanol by use of single molecular defined ruthenium and rhodium arene phosphine based catalysts.[127] Firstly, the [(η^6-*p*-cymene)RuH($PTAH^+$)X]/[RhH(η^5-C_5Me_5)($PTAH^+$)X] complex is formed *in situ* from the corresponding [Ru(η^6-*p*-cymene)-X_2(PTA)]/[Rh(η^5-C_5Me_5)X_2(PTA)] in a solution with acid.[127] The crucial role of the acid enabled an extremely facile formation of reactive Ru–H/Rh–H species.[127] When bis-(trifluoromethane)sulfonamide ($HNTf_2$) was employed as an acid additive for hydrogenation, a TON of 4752 was achieved using a [Ru(η^6-*p*-cymene)(PTA)] catalyst at 60 °C after 24 h.[127] Reaction of [(η^6-*p*-cymene)RuH($PTAH^+$)X]/[RhH(η^5-C_5Me_5)($PTAH^+$)X] with CO_2 afforded the κ^2-coordinated formate species $[(\eta^6$-*p*-cymene)$RuO_2CH(PTAH^+)]^+/[(\eta^5$-$C_5Me_5)RhO_2$-$CH(PTAH^+)]^+$.[127] Reaction with one equivalent of H_2 led to reduction beyond the formic acid stage to give the ruthenium/rhodium–hydroxymethanolate species, which is converted to the ruthenium/rhodium–methanolate complex *via* formation of intermediate formaldehyde and consumption of a second equivalent of H_2.[127] In the last step, hydrogenolysis of the Ru/Rh–OMe unit requires the third equivalent of H_2 to liberate methanol and completes the cycle by reforming the ruthenium/rhodium–hydride complex.[127]

A composite consisting of a ruthenium PNP pincer complex encapsulated in the MOF UiO-66 is used in tandem with the zirconium oxide nodes of UiO-66 and a ruthenium PNN pincer complex to hydrogenate carbon dioxide to methanol.[128] Reactions carried out in the presence of molecular sieves to remove water highlighted the beneficial effects of the ammonium functional group in UiO-66–NH_3^+ and resulted in a four-fold increase in activity.[128] As a result of the modular nature of the catalyst system, the highest reported turnover number (TON) (19 000) and turnover frequency (TOF) (9100 h^{-1}) for the hydrogenation of CO_2 to CH_3OH are obtained at 70 °C.[128] This reaction was readily recyclable, leading to a cumulative TON of 100 000 after 10 reaction cycles.[128]

A highly active and stable PdMo intermetallic catalyst, h-PdMo, has enabled room temperature methanol synthesis by hydrogenation of CO_2.[129] The h-PdMo catalyst can be prepared by the facile ammonolysis of an oxide precursor, and the catalyst exhibited long-term stability in air.[129] The methanol synthesis activity of the h-PdMo catalysts was further improved by pressurization, which resulted in continuous CO_2 hydrogenation to methanol at room temperature to afford a TOF of 0.15 h^{-1} at 0.9 MPa and 25 °C, which is the highest at 25 °C.[129]

Methanol can also be obtained by disproportionation of HCOOH with ruthenium(II) complexes supported by external phosphine ligands. As an alternative method to convert CO_2 to methanol without employing excess H_2, the catalytic disproportionation of formic acid to methanol and CO_2 has attracted much attention.[130–135] Different pathways have been proposed based on the possible organic intermediates involved in the disproportionation of formic acid, as shown in Scheme 7.9, where formic acid first undergoes a catalytic reduction to formaldehyde by transfer hydrogenation by a second equivalent of formic acid.[136] Formaldehyde is reduced to methanol by transfer hydrogenation with a third equivalent of formic acid.[136] In addition, methanol may be produced by hydrogenation of CO_2 from the competitive dehydrogenation of formic acid (Scheme 7.9).[136] Although the disproportionation reaction was efficient at 150 °C, it is significantly slowed down only at temperatures below 40 °C.[136] For example, the transformation of 2.4 mmol formic acid afforded methanol in 11.9% and 7.6% yield at 150°C and 80 °C, respectively, while the yield drops to 1.0% at 40 °C.[136]

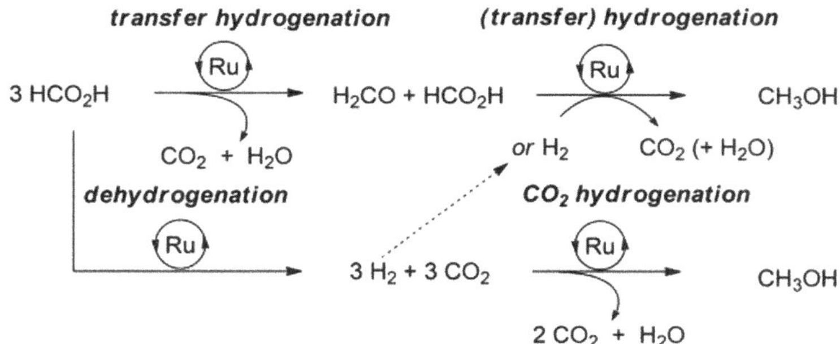

Scheme 7.9 Proposed pathways for the disproportionation of formic acid to methanol. Reproduced from ref. 136 with permission from John Wiley and Sons, Copyright 2014 Wiley-VCH Verlag GmbH & Co. KGaA, Weinheim.

Under optimized conditions, the disproportionation of formic acid was performed using 0.6 mol% [Ru(COD)(methylallyl)$_2$], $CH_3C(CH_2PPh_2)_3$ (triphos) and 1.5 mol% methanesulfonic acid (MSA) at 150 °C to afford 50.2% methanol yield, which is the best selectivity obtained from the disproportionation of formic acid to methnanol.[130–136]

7.6 Methane

Natural gas is mostly composed of methane (CH_4), accounting for 24% of the primary energy consumed in the world.[137] The current extraction of natural gas is primarily sourced from fossil fuels. However, extensive efforts have been devoted to sustainable production of methane by hydrogenation of CO_2 with heterogeneous catalysts.[138–142] In the case of homogeneous catalysis, a novel hydrogenation of CO_2 to CH_4 *via* CO and CH_3OH [eqn (7.13)] was reported to proceed in the presence of $Ru_3(CO)_{12}$–KI at a higher reaction temperature (240 °C) but under an almost equivalent pressure (90–140 atm).[143] In a typical experiment, an *N*-methylpyrrolidone (20 mL) solution of $Ru_3(CO)_{12}$ (0.1 mmol) and KI (10 mmol) was stirred in a 100 mL Hastelloy autoclave for over 1 h and a CO_2–H_2 (1 : 3) gas mixture was introduced under 80 atm at 30–32 °C.[143] The autoclave was then heated so that its interior temperature was raised and kept at 240 °C for 3 h.[143] GLC analyses of the resulting gas and solution revealed the formation of methanol (18.9 mmol or a 32 ton based on ruthenium atom), carbon monoxide (6.5 mmol), methane (4.7 mmol) and ethane (0.1 mmol).[143]

$$CO_2 + H_2 \xrightarrow[Ru_3(CO)_{12}\text{–KI}]{} CO + CH_3OH + CH_4 \tag{7.13}$$

The time course of the hydrogenation of CO_2 shows the successive formations of CO, CH_3OH and CH_4.[143] The yield of CO reached its maximum in the early stages of the reaction.[143] The yield of CH_3OH increased with a decrease in CO up to 3 h and then gradually decreased.[143] A constant increase in the yield of hydrocarbons may be due to the homogeneous hydrogenation of CH_3OH to CH_4, which was verified by the enhanced formation of CH_4 in a control experiment with additional methanol.[143] Scheme 7.10 shows a possible overall reaction mechanism of the catalytic reduction of CO_2 to CH_4. Although formation of an intermediate ruthenium formate complex was implied in the hydrogenation of CO_2 to alkyl formate catalysed by the ruthenium carbonyl complexes, the η^1-coordinated CO_2 complex may not be excluded from the mechanism of hydrogenation of CO_2 to CO.[143] Hydrogenation of CO to CH_3OH may be followed by further hydrogenation to CH_4, which is co-catalysed by an iodide anion.[143]

The hydrogenation of CO_2 to CH_4 at a low temperature has been achieved using the catalysis of [Ru]@P(IL^n-POSS)–Cl–A ($n = 2, 3, 4$) hybrids, wherein the molecular low-valent Ru complexes were encapsulated into the imidazolium chloride decorated porous poly(ionic liquid) skeleton.[144] [Ru]@P(ILn-POSS)–Cl–A catalysts were prepared *via* three steps of radical copolymerization of polyhedral oligomeric silsesquioxanes (POSS) with monomers of ionic bromide (VIm-n), anion exchange with NaCl and wet impregnation of $Ru_3(CO)_{12}$.[144] The hydrogenation of CO_2 to CH_4 was performed in the presence of [Ru]@P(ILn-POSS)–Cl–A at 160 °C under CO_2 (2 MPa) and H_2 (3 MPa) environments with the same dosage of Ru to afford CH_4 as the major product together with C_2H_6 as the side product.[144] The TON values toward CH_4 decreased in the order of [Ru]@P(IL^2-POSS)–Cl–A > [Ru]@P(IL^3-POSS)–Cl–A > [Ru]@P(IL^4-POSS)–Cl–A. The optimum TON value of 81 toward CH_4 was obtained on [Ru]@P(IL^2-POSS)–Cl–A after reaction for 16 h.[144]

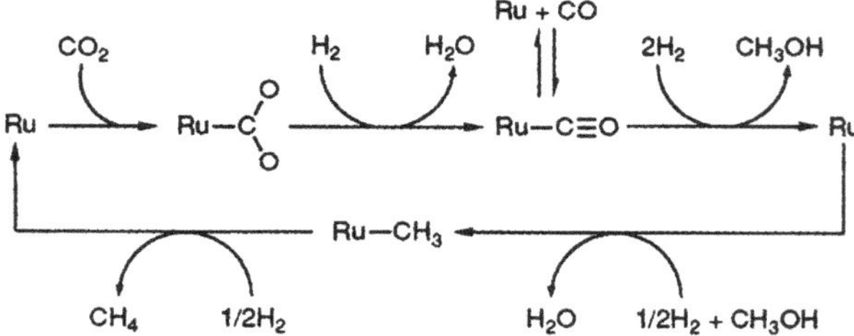

Scheme 7.10 A suggested reaction pathway for hydrogenation of CO_2 to CO, CH_3OH and CH_4 with a Ru complex. Reproduced from ref. 143 with permission from the Royal Society of Chemistry.

The recyclability of [Ru]@P(IL^2-POSS)–Cl–A was examined and the methanation of CO_2 for each run was conducted under standard conditions.[144] After the reaction, the employed catalyst was isolated and freeze-dried for the next run.[144] Overall five runs were performed, wherein the catalyst durability increased in an order of [Ru]@P(IL^2-POSS)–Cl–A < [Ru]@P(IL^3-POSS)–Cl–A < [Ru]@P(IL^4-POSS)–Cl–A.[144] In the catalytic mechanism of hydrogenation of CO_2 to CH_4 with [Ru]@P(IL^n-POSS)–Cl–A, $[Ru_3(Cl)(CO)_{12-m}]^-$ reacts with H_2 to form the Ru–H species with elimination of HCl.[144] Then, CO_2 is inserted into the Ru–H bond to form the formate complex, followed by stepwise hydrogenation to methanol and the Ru–methyl complex in sequence.[144] The final protonation with HCl affords CH_4 and regenerates the active Ru site.[144]

7.7 NADH

When HCOOH in Scheme 7.6 was replaced by 1,4-dihydro-β-nicotinamide adenine dinucleotide (NADH), H_2 was evolved from NADH in the presence of a catalytic amount of a water soluble [C,N] cyclometalated Ir complexes ($[M_1–OH_2]^+$) in an acidic aqueous solution (Scheme 7.11).[145] The H_2 evolution was accompanied by oxidation of NADH to the oxidized form, β-nicotinamide adenine dinucleotide (NAD^+), as confirmed by ^{1}H NMR.[145] The yield and turnover number (TON) reached up to 96% and 6.9 (20 min), respectively.[145] The reaction of $[M_1–OH_2]^+$ and the deprotonated complex ($[M_2–OH_2]^0$) with NADH afforded the corresponding anionic Ir-hydride complex as indicated by the negative-ion ESI mass spectrum ($m/z = 515.2$).[145] The reaction of the hydride complex with proton in water yields H_2, which is the rate-determining step in the catalytic cycle in Scheme 7.11 (top left-hand side) as indicated by the saturation behaviour of TOF with increasing the concentration of NADH.[145] Under basic conditions (*e.g.*, pH = 8), the reverse reaction of H_2 production from NADH, *i.e.*, hydrogenation of NAD^+ by H_2 with $[M_2–OH_2]^0$ occurred to produce NADH regioselectively (1,4-reduced form).[145] The yield based on the amount of NAD^+ and turnover number (TON) reached 97% and 9.3 (90 min), respectively.[145]

The TOF of H_2 evolution from NADH increased with a decrease in pH in the region between 4.1 and 7.0, which overlaps with the ratio of $[M_1–OH_2]^+$, whereas the pH dependence of TOF for hydrogenation of NAD^+ overlaps with the ratio of $[M_2–OH_2]^0$.[145] This indicates that $[M_1–OH_2]^+$ reacts with NADH to produce H_2 and that $[M_2–OH_2]^0$

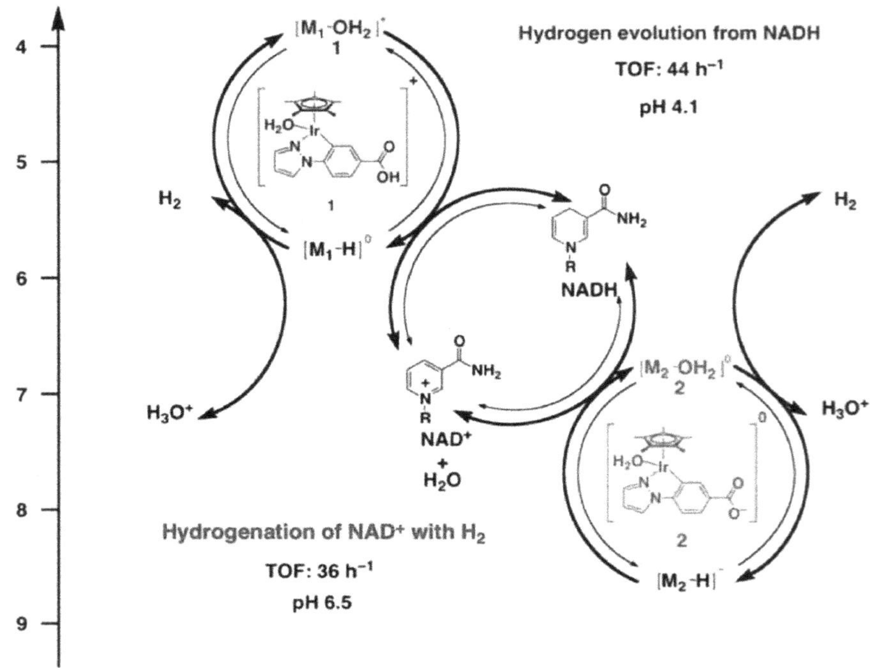

Scheme 7.11 Catalytic mechanism for interconversion between NADH and H_2 with a water soluble [C,N] cyclometalated Ir complexes (1 and 2) depending on pH. Adapted from ref. 145 with permission from American Chemical Society, Copyright 2012.

reacts with H_2 to reduce NAD^+ to NADH as the case of interconversion between HCOOH and H_2.[88] At pH 6.5, the TOF for the formation of NADH was maximized (36 h^{-1}), whereas the TOF for the H_2 evolution reached 44 h^{-1} at pH 4.1.[145]

The rate-determining step in the catalytic hydrogenation of NAD^+ with H_2 is the formation of the Ir–H complex, which reacts with NAD^+ rapidly to produce NADH (the right-hand catalytic cycle in Scheme 7.11).[145] Formation of the Ir–H complex under an atmospheric pressure of H_2 was confirmed by ESI mass spectrometry, ^{1}H NMR and UV-vis absorption spectra.[145] This is the first example of a hydrogenase functional mimic using a water-soluble iridium–aqua complex which can catalyse the oxidation of H_2 with NAD^+ to produce protons and NADH and also the reduction of protons with NADH to produce H_2 and NAD^+ in water under atmospheric pressure at room temperature.[145] In such a case, NADH is regarded as a solid hydrogen carrier that is soluble in water.

7.8 Reduction of CO_2 by Enzymes

NADH can reduce CO_2 to methanol by using three dehydrogenases, as shown in Scheme 7.12, where the formate dehydrogenase (FateDH) converts gaseous CO_2 into formate, then the formaldehyde dehydrogenase (FaldDH) converts formate into formaldehyde and finally alcohol dehydrogenase (YADH) converts formaldehyde into methanol.[146] Overall three equivalents of NADH are required to reduce CO_2 to methanol.[146] The optimum ratio of the three polyenzymatic systems FateDH, FaldH and YADH was found to be 0.01, 0.15 and 0.75 $g L^{-1}$ of commercial enzymatic powder, respectively.[146] Immobilization of the enzymes is a critical concern for biochemical processes as it provides not only stabilization and easier use, but also improved activity.[147–149]

The polyenzymatic system was encapsulated by an optimized silica sol–gel technique using natural phospholipids (egg lecithin) and lactose to protect the enzymes leading to phospholipids–silica nanocapsules (NPS).[146] By using an efficient enzymatic encapsulation host such as phospholipids–silica nanocapsules (NPS), the activity of the polyenzymatic system was increased by a factor close to 30, leading to a productivity in methanol (in static at 5 bar for 3 h) of 4.3 $mmol\, g_{commercial\ enzymatic\ powder}{}^{-1}$ equivalent to what was previously reported using a continuous CO_2 bubbling.[146] The productivity in methanol per mass of catalyst corresponds to 42 μmol methanol produced per g_{NPS} in 3 h.[146]

Three different methods (co-encapsulation into Ca–alginate beads, co-absorption onto zirconium(IV) phosphate (ZrP) and covalent binding to dialdehydecellulose [DAC]) of co-immobilization of the three dehydrogenases, FateDH, FaldDH, and alcohol dehydrogenase, were tested to examine at what extension the support and the

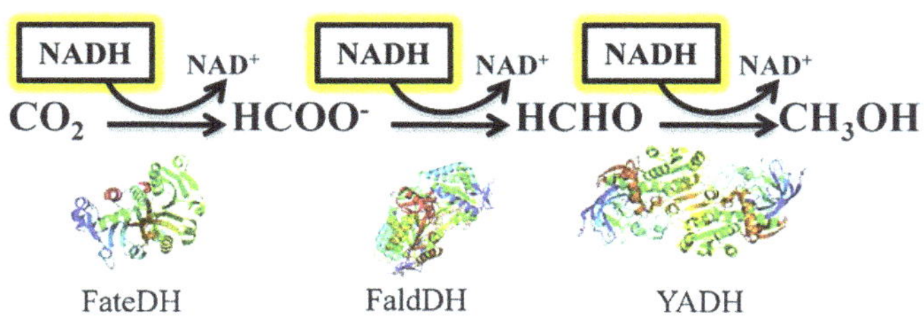

Scheme 7.12 Reduction scheme of CO_2 to methanol by three dehydrogenase enzymes (FateDH, FaldDH and YADH). Reproduced from ref. 146 with permission from the Royal Society of Chemistry.

immobilization method can influence the activity of the enzymatic pool.[147] DAC was found to be the best method of immobilization, allowing longer life on enzymes and repeated recycling of the supported enzymes, increasing the overall methanol production with respect to the free enzymes.[147] Free enzymes have a limited lifetime (after 3 h they are inactive). In contrast to free enzymes, the DAC immobilized enzymes can be recycled for at least five cycles, for a total lifetime of over 15 h.[147]

Coupling CO_2 and ethanol into L-lactic acid was also achieved by using a simple multienzyme route, developing a novel and sustainable alternative way to synthesize building blocks for biodegradable polymers.[149,150] The catalytic process consisted of three single-enzyme routes: (1) the oxidation of ethanol to acetaldehyde by alcohol dehydrogenase (ADH) in the presence of NAD^+, (2) the synthesis of pyruvate from CO_2 and acetaldehyde catalysed by pyruvate decarboxylase and (3) the reduction of pyruvate to L-lactic acid by LDH in the presence of NADH.[149,150] The cycling of NAD^+/NADH could be achieved *via* the first and third single-enzyme route.[150] Using this approach in a batch process and with continuous feeding of ethanol, 41% of ethanol was converted into lactate after 4 days.[150]

A dual enzyme cascade system was also developed to synthesize (*S*)-1-(2-chlorophenyl)ethanol from *o*-chloroacetophenone by combining an alcohol dehydrogenase (ADH) and aldo-keto reductase (AKR), which were sequentially anchored on the surface of porous P(GMA-EDMA) organic polymer microspheres.[151] The two enzymes were combined through the spontaneous cross-linking of the SpyTag/SpyCatcher system and then immobilized *via* a click reaction between the inserted ncAA-bearing azide groups and the vector functionalized with alkyne groups.[151] (*S*)-1-(2-Chlorophenyl)ethanol is a crucial chiral building block for pharmaceuticals and is synthesized through the asymmetric reduction of *o*-chloroacetophenone by AKR. The disorderly co-immobilized dual-enzyme resulted in a 41.6% conversion rate and an enantiomeric excess (e.e.) of 98.6%, whereas the (*S*)-1-(2-chlorophenyl) ethanol yield obtained by the ordered dual-enzyme coatings increased to 73.7% with e.e. values above 99.9%.[151]

References

1. W. Scholz, F. Galván and F. F. de la Rosa, *Sol. Energy Mater. Sol. Cells*, 1995, **39**, 61–69.
2. F. F. de la Rosa, O. Montes and F. Galván, *Biotechnol. Bioeng.*, 2001, **74**, 539–543.

3. M. Mubarakshina, S. Khorobrykh and B. Ivanov, *Biochim. Biophys. Acta*, 2006, **1757**, 1496–1503.
4. R. Simard, *Eng. J.*, 1948, **31**, 219–225.
5. E. Burgess, *J. Am. Rocket Soc.*, 1946, **65**, 18–19.
6. S. Bonifacio, G. Festa and A. R. Sorge, *J. Propul. Power*, 2013, **29**, 1130–1137.
7. H. Wang, G. Zhao and M. Pumera, *J. Am. Chem. Soc.*, 2014, **136**, 2719–2722.
8. A. Martín, B. Jurado-Sánchez, A. Escarpaand and J. Wang, *Small*, 2015, **11**, 499–506.
9. H. Wang, Z. Sofer, A. Y. S. Eng and M. Pumera, *Chem. – Eur. J.*, 2014, **20**, 14946–14950.
10. S. Fukuzumi, *Joule*, 2017, **1**, 689–738.
11. S. Fukuzumi and Y. Yamada, *ChemElectroChem*, 2016, **3**, 1978–1989.
12. O. C. Esan, X. Shi, Z. Pan, Y. Liu, X. Huo, L. An and T. S. Zhao, *J. Power Sources*, 2022, **548**, 232114.
13. N. Nelson and C. F. Yocum, *Annu. Rev. Plant Biol.*, 2006, **57**, 521–565.
14. L. Gurrieri, S. Fermani, M. Zaffagnini, F. Sparla and P. Trost, *Trends Plant Sci.*, 2021, **26**, 898–912.
15. A. Mehler, *Arch. Biochem. Biophys.*, 1951, **34**, 339–351.
16. C. Miyake, *Plant Cell Physiol.*, 2010, **51**, 1951–1963.
17. K. Asada, *Philos. Trans. R. Soc. London, Ser. B*, 2000, **355**, 1419–1431.
18. T. Roach, C. S. Na and A. Krieger-Liszkay, *Plant J.*, 2015, **81**, 759–766.
19. M. M. Borisova, M. A. Kozuleva, N. N. Rudenko, I. A. Naydov, I. B. Klenina and B. N. Ivanov, *Biochim. Biophys. Acta, Bioenerg.*, 2012, **1817**, 1314–1321.
20. S. Takahashi and N. Murata, *Trends Plant Sci.*, 2008, **13**, 178–182.
21. M. Pilon, K. Ravetand and W. Tapken, *Biochim. Biophys. Acta, Bioenerg.*, 2011, **1807**, 989–998.
22. Y. Higashi, K. Takechi, H. Takano and S. Takio, *Am. J. Plant Sci.*, 2015, **6**, 591–601.
23. M. R. Badger, S. von Caemmerer, S. Ruuska and H. Nakano, *Philos. Trans. R. Soc. London, Ser. B*, 2000, **355**, 1433–1446.
24. I. V. Kuvykin, A. V. Vershubskii, V. V. Ptushenko and A. N. Tikhonov, *Biochemistry*, 2008, **73**, 1063–1075.
25. H. Sies, *J. Biol. Chem.*, 2014, **289**, 8735–8741.
26. J. R. Stone and S. Yang, *Antioxid. Redox Signaling*, 2006, **8**, 243–270.
27. G. P. Bienert, J. K. Schjoerringand and T. P. Jahn, *Biochim. Biophys. Acta*, 2006, **1758**, 994–1003.
28. E. A. Veal, A. M. Day and B. A. Morgan, *Mol. Cell*, 2007, **26**, 1–14.
29. S. Fukuzumi, Y. Yamada and K. D. Karlin, *Electrochim. Acta*, 2012, **82**, 493–511.
30. S. Fukuzumi and Y. Yamada, *Aust. J. Chem.*, 2014, **67**, 354–364.
31. X. Fang, X. Huang, Q. Hu, B. Li, C. Hu, B. Ma and Y. Ding, *Chem. Commun.*, 2024, **60**, 5354–5368.
32. S. Fukuzumi, *Biochim. Biophys. Acta, Bioenerg.*, 2016, **1857**, 604–611.
33. S. Fukuzumi, Y.-M. Lee and W. Nam, *Chem. – Eur. J*, 2018, **24**, 5016–5031.
34. W. Fang and L. Wang, *Catalysts*, 2023, **13**, 1325.
35. S. Fukuzumi, Y.-M. Lee and W. Nam, *Chin. J. Catal.*, 2021, **42**, 1241–1252.
36. S. Wei, S. Chang, H. Li, Z. Fang, L. Zhu and Y. Xu, *Green Chem.*, 2024, **26**, 6382–6403.
37. Y. H. Hong, Y.-M. Lee, W. Nam and S. Fukuzumi, *J. Porphyrins phthalocyanines*, 2023, **27**, 11–22.
38. D. Tan, R. Zhuang, R. Chen, M. Ban, W. Feng, F. Xu, X. Chen and Q. Wang, *Adv. Funct. Mater.*, 2024, **34**, 2311655.
39. J. M. Campos-Martin, G. Blanco-Brieva and J. L. G. Fierro, *Angew. Chem., Int. Ed.*, 2006, **45**, 6962–6984.
40. A. A. Ingle, S. Z. Ansari, D. Z. Shende, K. L. Wasewar and A. B. Pandit, *Environ. Sci. Pollut. Res. Int.*, 2022, **29**, 86468–86484.

41. J. García-Serna, T. Moreno, P. Biasi, M. J. Cocero, J.-P. Mikkolab and T. O. Salmi, *Green Chem.*, 2014, **16**, 2320–2343.
42. C. Samanta, *Appl. Catal., A*, 2008, **350**, 133–149.
43. J. K. Edeards, S. J. Freakley, A. F. Carley, C. J. Kiely and G. J. Hutchings, *Acc. Chem. Res.*, 2014, **47**, 845–854.
44. J. K. Edwards, S. J. Freakley, R. J. Lewis, J. C. Pritchard and G. J. Hutchings, *Catal. Today*, 2015, **248**, 3–9.
45. J. K. Edwards, B. Solsona, E. Ntainjua, N. A. F. Carley, A. A. Herzing, C. J. Kielyand and G. J. Hutchings, *Science*, 2009, **323**, 1037–1041.
46. S. Shibata, T. Suenobu and S. Fukuzumi, *Angew. Chem., Int. Ed.*, 2013, **52**, 12327–12331.
47. Y. Maenaka, T. Suenobu and S. Fukuzumi, *J. Am. Chem. Soc.*, 2012, **134**, 367–374.
48. S. Shibata, T. Suenobu and S. Fukuzumi, *Angew. Chem., Int. Ed.*, 2013, **52**, 12327–12331.
49. S. Fukuzumi, S. Kuroda and T. Tanaka, *J. Am. Chem. Soc.*, 1985, **107**, 3020–3027.
50. S. Kakuda, C. J. Rolle, K. Ohkubo, M. A. Siegler, K. D. Karlin and S. Fukuzumi, *J. Am. Chem. Soc.*, 2015, **137**, 3330–3337.
51. S. Ogo, T. Yatabe, T. Tome, R. Takenaka, Y. Shiota and K. Kato, *J. Am. Chem. Soc.*, 2023, **145**, 4384–4388.
52. S. Kato, J. Jung, T. Suenobu and S. Fukuzumi, *Energy Environ. Sci.*, 2013, **6**, 3756–3764.
53. A. Das, V. Joshi, D. Kotkar, V. S. Pathak, V. Swayambunathan, P. V. Kamat and P. K. Ghosh, *J. Phys. Chem. A*, 2001, **105**, 6945–6954.
54. S. Fukuzumi and K. Ohkubo, *Chem. – Eur. J.*, 2000, **6**, 4532–4535.
55. S. Fukuzumi and K. Ohkubo, *J. Am. Chem. Soc.*, 2002, **124**, 10270–10271.
56. S. Fukuzumi, M. Patz, T. Suenobu, Y. Kuwahara and S. Itoh, *J. Am. Chem. Soc.*, 1999, **121**, 1605–1606.
57. A. Melis, *Plant Sci.*, 2009, **177**, 272–280.
58. Y. Isaka, S. Kato, D. Hong, T. Suenobu, Y. Yamada and S. Fukuzumi, *J. Mater. Chem. A*, 2015, **3**, 12404–12412.
59. Y. Shiraishi, S. Kanazawa, Y. Kofuji, H. Sakamoto, S. Ichikawa, S. Tanaka and T. Hirai, *Angew. Chem., Int. Ed.*, 2014, **53**, 13454–13459.
60. Y. Kofuji, Y. Isobe, Y. Shiraishi, H. Sakamoto, S. Tanaka, S. Ichikawa and T. Hirai, *J. Am. Chem. Soc.*, 2016, **138**, 10019–10025.
61. Y. Shiraishi, T. Takii, T. Hagi, S. Mori, Y. Kofuji, Y. Kitagawa, S. Tanaka, S. Ichikawa and T. Hirai, *Nat. Mater.*, 2019, **18**, 985–993.
62. R. Ma, L. Wang, H. Wang, Z. Liu, M. Xing, L. Zhu, X. Meng and F.-S. Xiao, *Appl. Catal., B*, 2019, **244**, 594–603.
63. X. Zhang, P. Ma, C. Wang, L. Y. Gan, X. Chen, P. Zhang, Y. Wang, H. Li, L. Wang, X. Zhou and K. Zheng, *Energy Environ. Sci.*, 2022, **15**, 830–842.
64. X. Xu, Y. Sui, W. Chen, W. Huang, X. Li, Y. Li, D. Liu, S. Gao, W. Wu, C. Pan, H. Zhong, H.-R. Wen and M. Wen, *Appl. Catal., B*, 2024, **341**, 123271.
65. Y. Zhang, C. Pan, G. Bian, J. Xu, Y. Dong, Y. Zhang, Y. Lou, W. Liu and Y. Zhu, *Nat. Energy*, 2023, **8**, 361–371.
66. K. Mase, M. Yoneda, Y. Yamada and S. Fukuzumi, *Nat. Commun.*, 2016, **7**, 11470.
67. K. Mase, K. Ohkubo and S. Fukuzumi, *J. Am. Chem. Soc.*, 2013, **135**, 2800–2808.
68. Z. Chen, J. J. Concepcion, N. Song and T. J. Meyer, *Chem. Commun.*, 2014, **50**, 8053–8056.
69. K. Mase, M. Yoneda, Y. Yamada and S. Fukuzumi, *ACS Energy Lett.*, 2016, **1**, 913–919.
70. S. Fukuzumi, Y. Yamada and K. D. Karlin, *Electrochim. Acta*, 2012, **82**, 493–511.
71. S. Fukuzumi and Y. Yamada, *Aust. J. Chem.*, 2014, **67**, 354–364.

72. S. Yamazaki, Z. Siroma, H. Senoh, T. Ioroi, N. Fujiwara and K. Yasuda, *J. Power Sources*, 2008, **178**, 20–25.
73. Y. Yamada, Y. Fukunishi, S. Yamazaki and S. Fukuzumi, *Chem. Commun.*, 2010, **46**, 7334–7336.
74. Y. Yamada, S. Yoshida, T. Honda and S. Fukuzumi, *Energy Environ. Sci.*, 2011, **4**, 2822–2825.
75. Y. Yamada, M. Yoneda and S. Fukuzumi, *Chem. – Eur. J.*, 2013, **19**, 11733–11741.
76. Y. Yamada, M. Yoneda and S. Fukuzumi, *Energy Environ. Sci.*, 2015, **8**, 1698–1701.
77. S. Fukuzumi, *Eur. J. Inorg. Chem.*, 2008, 1351–1362.
78. D. Mellmann, P. Sponholz, H. Junge and M. Beller, *Chem. Soc. Rev.*, 2016, **45**, 3954–3988.
79. M. Lie, Y. Xu, Y. Meng, L. Wang, H. Wang, Y. Huang, N. Onishi, L. Wang, Z. Fan and Y. Himeda, *Adv. Energy Mater.*, 2022, **12**, 2200817.
80. S. Fukuzumi and T. Suenobu, *Dalton Trans.*, 2013, **42**, 18–28.
81. N. Onishi, R. Kanega, H. Kawanami and Y. Himeda, *Molecules*, 2022, **27**, 455.
82. Y. H. Hong, Y.-M. Lee, W. Nam and S. Fukuzumi, *Inorg. Chem. Front.*, 2024, **11**, 981–997.
83. W.-H. Wang, Y. Himeda, J. T. Muckerman, G. F. Manbeck and E. Fujita, *Chem. Rev.*, 2015, **115**, 12936–12973.
84. H. Hayashi, S. Ogo and S. Fukuzumi, *Chem. Commun.*, 2004, 2714–2715.
85. S. Ogo, R. Kabe, H. Hayashi, R. Harada and S. Fukuzumi, *Dalton Trans.*, 2006, 4657–4663.
86. H. Hayashi, S. Ogo, T. Abura and S. Fukuzumi, *J. Am. Chem. Soc.*, 2003, **125**, 14266–14267.
87. S. Fukuzumi, T. Suenobuand and Y. Maenaka, *JP pat.*, JP2010048477A, 2010, p. 03004.
88. Y. Maenaka, T. Suenobu and S. Fukuzumi, *Energy Environ. Sci.*, 2012, **5**, 7360–7367.
89. J. F. Hull, Y. Himeda, W.-H. Wang, B. Hashiguchi, R. Periana, D. J. Szalda, J. T. Muckerman and E. Fujita, *Nat. Chem.*, 2012, **4**, 383–388.
90. C. A. Huff and M. S. Sanford, *ACS Catal.*, 2013, **3**, 2412–2416.
91. Z. Wang, S.-M. Lu, J. Li, J. Wang and C. Li, *Chem. – Eur. J.*, 2015, **21**, 12592–12595.
92. J. Eppinger and K. W. Huang, Formic Acid as a Hydrogen Energy Carrier, *ACS Energy Lett.*, 2017, **2**, 188–195.
93. S. Chatterjee, I. Dutta, Y. Lum, Z. Lai and K.-W. Huang, *Energy Environ. Sci.*, 2021, **14**, 1194–1246.
94. R. Verron, E. Puig, P. Sutra, A. Igau and C. Fischmeister, *ACS Catal.*, 2023, **13**, 5787–5794.
95. D. Wei, R. Sang, P. Sponholz, H. Junge and M. Beller, *Nat. Energy*, 2022, **7**, 438–447.
96. W. Ma, W. Xiong, J. Hu, J. Geng and X. Hu, *Green Chem.*, 2024, **26**, 4192–4198.
97. G. Ji, C. Li, B. Fan, G. Wang, Z. Sun, M. Jiang, L. Ma, L. Yan and Y. Ding, *ACS Catal.*, 2024, **14**, 1595–1607.
98. S. Siangwata, A. Hamilton, G. J. Tizzard, S. J. Coles and G. R. Owen, *ChemCatChem*, 2024, **16**, e202301627.
99. K. Tsuchiya, J.-D. Huangand and K.-I. Tominaga, *ACS Catal.*, 2013, **3**, 2865–2868.
100. Q. Chen, C. Shen, G. Zhu, X. Zhang, C.-L. Lv, B. Zeng, S. Wang, J. Li, W. Fan and L. He, *ACS Catal.*, 2021, **11**, 9390–9396.
101. Q. Chen, X. Kang, X. Zhang, Y. Cao and L. He, *J. Org. Chem.*, 2023, **88**, 5044–5051.
102. A. M. Bahmanpour, A. Hoadley and A. Tanksale, *Rev. Chem. Eng.*, 2014, **30**, 583–604.
103. L. E. Heim, H. Konnerth and M. H. G. Prechtl, *Green Chem.*, 2017, **19**, 2347–2355.
104. S. Zhao, H.-Q. Liang, X.-M. Hu, S. Li and K. Daasbjerg, *Angew. Chem., Int. Ed.*, 2022, **61**, e202204008.

105. L. E. Heim, H. Konnerth and M. H. G. Prechtl, *ChemSusChem*, 2016, **9**, 2905–2907.
106. L. Deng, Z. Wang, X. Jiang, J. Xu, Z. Zhou, X. Li, Z. You, M. Ding, T. Shishido, X. Liu and M. Xu, *Appl. Catal., B*, 2023, **322**, 122124.
107. A. M. Bahmanpour, A. Hoadley and A. Tanksale, *Green Chem.*, 2015, **17**, 3500–3507.
108. A. M. Bahmanpour, A. Hoadley, S. H. Mushrif and A. Tanksale, *ACS Sustainable Chem. Eng.*, 2016, **4**, 3970–3977.
109. L. Wang, M. Z. Ertem, R. Kanega, K. Murata, D. J. Szalda, J. T. Muckerman, E. Fujita and Y. Himeda, *ACS Catal.*, 2018, **8**, 8600–8605.
110. S. Patra, A. Kumar and S. K. Singh, *Inorg. Chem.*, 2022, **61**, 4618–4626.
111. T. Suenobu, Y. Isaka, S. Shibata and S. Fukuzumi, *Chem. Commun.*, 2015, **51**, 1670–1672.
112. J. Parthiban, M. K. Awasthi, T. A. Kharde, K. Kalita and S. K. Singh, *Dalton Trans.*, 2024, **53**, 4363–4389.
113. L. E. Heim, N. E. Schlorer, J. H. Choi and M. H. Prechtl, *Nat. Commun.*, 2014, **5**, 3621.
114. N. Lu, X. Yan, H. L. Tan, H. Kobayashi, X. Yu, Y. Li, J. Zhang, Z. Peng, J. Sui, Z. Zhang, W. Liu, R. Li and B. Li, *Int. J. Hydrogen Energy*, 2022, **47**, 27877–27886.
115. Y. Sun, Y. Xiao, L. Ren, Z. Cheng, Y. Niu, Z. Li and S. Zhang, *J. Phys. Chem. Lett.*, 2024, **15**, 4538–4545.
116. J. Zhong, X. Yang, Z. Wu, B. Liang, Y. Huang and T. Zhang, *Chem. Soc. Rev.*, 2020, **49**, 1385–1413.
117. Y. Xu, L. Wang, Q. Zhou, Y. Li, L. Liu, W. Nie, R. Xu, J. Zhang, Z. Cheng, H. Wang, Y. Huang, T. Wei, Z. Fan and L. Wang, *Coord. Chem. Rev.*, 2024, **508**, 215775.
118. G. Pacchioni, *ACS Catal.*, 2024, **14**, 2730–2745.
119. G. Xie, X. Bai, F. Yu, Q. Yang and Z.-J. Wang, *Catal. Today*, 2024, **434**, 114702.
120. R. Sen, A. Goeppert and G. K. S. Prakash, *Angew. Chem., Int. Ed.*, 2022, **61**, e202207278.
121. S. Kostera and L. Gonsalvi, *ChemCatChem*, 2024, **16**, e202301391.
122. N. Onishi and Y. Himeda, *Coord. Chem. Rev.*, 2022, **472**, 214767.
123. S.-T. Bai, G. De Smet, Y. Liao, R. Sun, C. Zhou, M. Beller, B. U. W. Maes and B. F. Sels, *Chem. Soc. Rev.*, 2021, **50**, 4259–4298.
124. N. Onishi and Y. Himeda, *Chem. Catal.*, 2022, **2**, 242–252.
125. S. Xie, W. Zhang, X. Lan and H. Lin, *ChemSusChem*, 2020, **13**, 6141–6159.
126. R. Kanega, N. Onishi, S. Tanaka, H. Kishimoto and Y. Himeda, *J. Am. Chem. Soc.*, 2021, **143**, 1570–1576.
127. M. Trivedi, P. Sharma, I. K. Pandey, A. Kumar, S. Kumar and N. P. Rath, *Chem. Commun.*, 2021, **57**, 8941–8944.
128. T. M. Rayder, A. T. Bensalah, B. Li, J. A. Byers and C.-K. Tsung, *J. Am. Chem. Soc.*, 2021, **143**, 1630–1640.
129. H. Sugiyama, M. Miyazaki, M. Sasase, M. Kitano and H. Hosono, *J. Am. Chem. Soc.*, 2023, **145**, 9410–9416.
130. A. J. M. Miller, D. M. Heinekey, J. M. Mayer and K. I. Goldberg, *Angew. Chem., Int. Ed.*, 2013, **52**, 3981–3984.
131. M. C. Neary and G. Parkin, *Chem. Sci.*, 2015, **6**, 1859–1865.
132. K. Sordakis, A. Tsurusaki, M. Iguchi, H. Kawanami, Y. Himeda and G. Laurenczy, *Green Chem.*, 2017, **19**, 2371–2378.
133. C. Chauvier, A. Imberdis, P. Thuéry and T. Cantat, *Angew. Chem., Int. Ed.*, 2020, **59**, 14019–14023.
134. E. Alberico, T. Leischner, H. Junge, A. Kammer, R. Sang, J. Seifert, W. Baumann, A. Spannenberg, K. Junge and M. Beller, *Chem. Sci.*, 2021, **12**, 13101–13119.
135. H. Fujita, S. Takemoto and H. Matsuzaka, *ACS Catal.*, 2021, **11**, 7460–7466.

136. S. Savourey, G. Lefèvre, J.-C. Berthet, P. Thuéry, C. Genre and T. Cantat, *Angew. Chem., Int. Ed.*, 2014, **53**, 10466–10470.
137. BP Statistical review of world energy, 2023.
138. M. A. A. Aziz, A. A. Jalil, S. Triwahyono and A. Ahmad, *Green Chem.*, 2015, **17**, 2647–2663.
139. M. A. Memon, Y. Jiang, M. A. Hassan, M. Ajmal, H. Wang and Y. Liu, *Catalysts*, 2023, **13**, 1514.
140. W. K. Fan and M. Tahi, *J. Environ. Chem. Eng.*, 2021, **9**, 105460.
141. Y. Cui, S. He, J. Yang, R. Gao, K. Hu, X. Chen, L. Xu, C. Deng, C. Lin, S. Peng and C. Zhang, *Molecules*, 2024, **29**, 374.
142. N. Albeladi, Q. A. Alsulami and K. Narasimharao, *Catalysts*, 2023, **13**, 1104.
143. K. Tominaga, Y. Sasaki, M. Kawai, T. Watanabe and M. Saito, *J. Chem. Soc., Chem. Commun.*, 1993, 629–631.
144. R. Wang, Y.-R. Du, G.-R. Ding, R. Zhang, P.-X. Guan and B.-H. Xu, *ACS Sustainable Chem. Eng.*, 2022, **10**, 5363–5373.
145. Y. Maenaka, T. Suenobu and S. Fukuzumi, *J. Am. Chem. Soc.*, 2012, **134**, 367–374.
146. R. Cazelles, J. Drone, F. Fajula, O. Ersen, S. Moldovan and A. Galarneau, *New J. Chem.*, 2013, **37**, 3721–3730.
147. C. Di Spiridione, M. Aresta and A. Dibenedetto, *Adv. Energy Sustainability Res.*, 2024, 2400081.
148. M. E. Aguirre and C. L. Ramírez, *Chem. - Eur. J.*, 2023, **29**, e202301113.
149. Y. Shu, W. Liang and J. Huang, *Green Chem.*, 2023, **25**, 4196–4221.
150. X. Tong, B. El-Zahab, X. Zhao, Y. Liu and P. Wang, *Biotechnol. Bioeng.*, 2011, **108**, 465–469.
151. Z. Luo, L. Qiao, H. Chen, Z. Mao, S. Wu, B. Ma, T. Xie, A. Wang, X. Pei and R. A. Sheldon, *Angew. Chem., Int. Ed.*, 2024, **63**, e202403539.

8 Conclusion and Perspective

Simple organic dyads have been shown to undergo short range charge separation (or charge shift) but tremendously slow charge recombination (or BET) with minimized energy loss, while long-range charge separation in the multistep ET processes requires a significant amount of energy loss in the natural photosynthetic reaction centre as well as in the model compounds. These simple donor–acceptor linked molecules have been successfully applied to build efficient photocatalytic systems in a variety of organic transformations, including as functional models of PSI, where plastoquinol analogues are oxidized to the corresponding plastoquinone analogues, accompanied by H_2 evolution (Scheme 4.1) and regioselective reduction of $NAD(P)^+$ to NAD(P)H [eqn (4.4)]. On the other hand, plastoquinone analogues are reduced to plastoquinol analogues by water to evolve O_2 with $[Fe^{II}(N4Py)]^{2+}$ *via* successive ET from $[Fe^{II}(N4Py)]^{2+}$ to the triplet excited state of plastoquinone analogues to produce $[Fe^{II}(O)(N4Py)]^{3+}$ which oxidizes water to evolve O_2 [eqn (5.2), PSII model]. The functional molecular models of PSI were then successfully combined with a functional molecular model of PSII to achieve production of H_2 (Scheme 6.1) and NAD(P)H (Scheme 6.2) by photocatalytic reduction of water and $NAD(P)^+$ using a homogeneous molecular photocatalyst and a reduction catalyst. Such functional molecular models of the combination of PSI and PSII provide valuable mechanistic insights enabling detection of reaction intermediates that would otherwise be impossible to achieve as long as heterogeneous photocatalysts are employed.

RSC Foundations No. 1
Artificial Photosynthesis
By Shunichi Fukuzumi

Published by the Royal Society of Chemistry, www.rsc.org

The molecular photocatalytic H_2 production from H_2O can be applied for the production of liquid solar fuels from water and CO_2 by replacing H_2 evolution catalysts by water and CO_2 reduction catalysts (Scheme 8.1). As described above, there have been extensive studies on catalytic hydrogenation of CO_2 to HCOOH, CO, HCHO, CH_3OH and CH_4. Selective production of each CO_2 species has been made possible by choosing an appropriate catalyst under proper experimental conditions. In many cases, however, a high pressure of H_2 is required to hydrogenate CO_2. An interesting alternative way is catalytic disproportionation of HCOOH to produce CH_3OH (Scheme 8.1). The photocatalytic regioselective reduction of $NAD(P)^+$ by water to produce NAD(P)H can also be combined with NAD(P)H dependent enzymatic reactions using water as an electron and proton source (Scheme 8.2). Thus, combination of PSI and PSII models (Scheme 6.2) with an immobilized polyenzymatic system is expected to make it

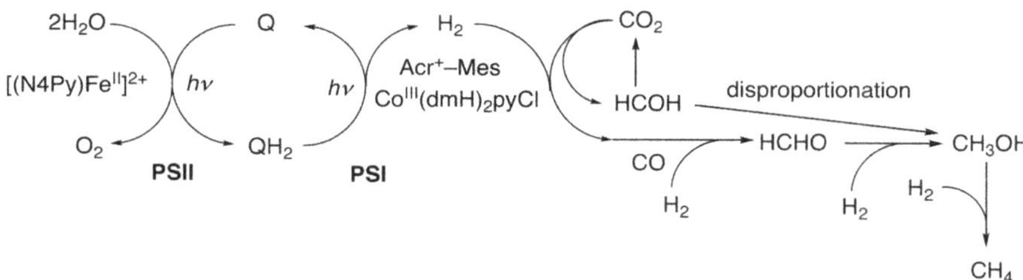

Scheme 8.1 Combination of models of PSI and PSII with catalytic hydrogenation of CO_2 to produce HCOOH, CO, HCHO, CH_3OH and CH_4 depending on hydrogenation catalysts using H_2O as an electron and proton source. The catalytic disproportionation of HCOOH affords CH_3OH.

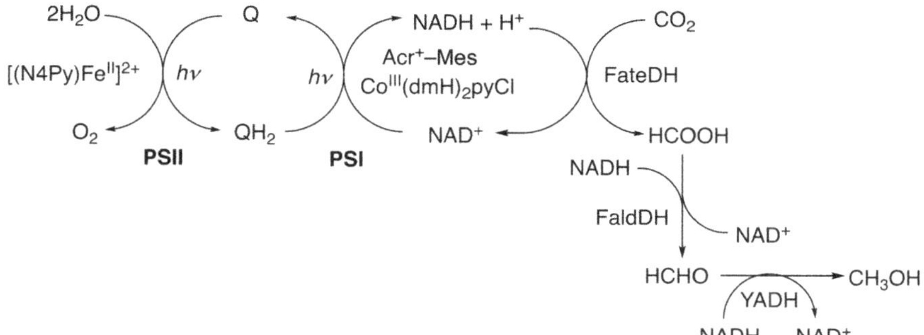

Scheme 8.2 Combination of models of PSI and PSII with enzymatic cascade hydrogenation of CO_2 to produce HCOOH, CO, HCHO, CH_3OH and CH_4 depending on hydrogenation catalyst.

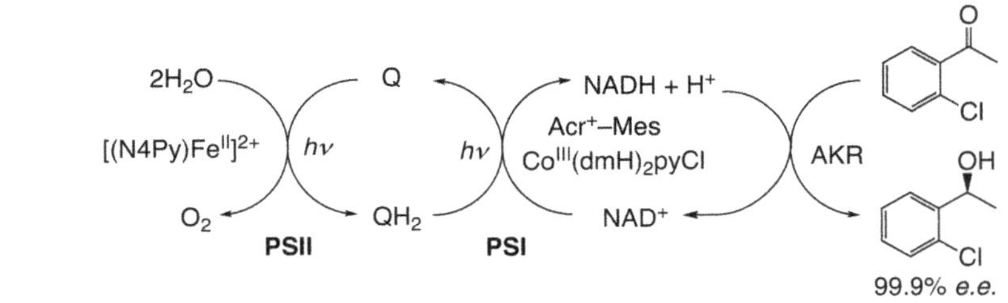

Scheme 8.3 Combination of models of PSI and PSII combined with NADH dependent enzymatic reactions for enantioselective reduction using water as an electron and proton source.

possible to produce methanol from CO_2 and water using solar energy, which has yet to be achieved.

The combination of PSI and PSII models (Scheme 6.2) with the immobilized polyenzymatic aldo-keto reductase (AKR) may also enable asymmetric reduction of *o*-chloroacetophenone by AKR to produce (*S*)-1-(2-chlorophenyl) ethanol with e.e. values above 99.9% using water as an electron and proton source (Scheme 8.3). Any NADH dependent enzyme can also be combined with the molecular modes of PSI and PSII to use water as an electron and proton source to achieve the NADH/NAD^+ cycle for the enzymatic reaction. Thus, molecular models of combination of PSI and PSII described in this book provide a promising perspective for production of solar fuels, reduction of CO_2 to value-added chemicals and NADH dependent enzymatic reactions using water as an electron and proton souse without consuming precious NADH.

www.ingramcontent.com/pod-product-compliance
Lightning Source LLC
Chambersburg PA
CBHW040857210326
41597CB00029B/4882

9781837071906